Enhancing Farm-Lev
Making through Inn

Enhancing Farm-Level Decision Making through Innovation

Editor

Matt J. Bell

MDPI • Basel • Beijing • Wuhan • Barcelona • Belgrade • Manchester • Tokyo • Cluj • Tianjin

Editor
Matt J. Bell
Hartpury University
UK

Editorial Office
MDPI
St. Alban-Anlage 66
4052 Basel, Switzerland

This is a reprint of articles from the Special Issue published online in the open access journal *Agriculture* (ISSN 2077-0472) (available at: https://www.mdpi.com/journal/agriculture/special_issues/farm_decision).

For citation purposes, cite each article independently as indicated on the article page online and as indicated below:

LastName, A.A.; LastName, B.B.; LastName, C.C. Article Title. *Journal Name* **Year**, *Volume Number*, Page Range.

ISBN 978-3-0365-3355-1 (Hbk)
ISBN 978-3-0365-3356-8 (PDF)

Cover image courtesy of Matt J. Bell

Contents

About the Editor

Matt J. Bell is an expert in agricultural systems and sustainable food production, with an interest in the complex relationship between animals, plants, soil, nutrients, water and climate. There has never been a greater need for solutions to enhance the way we produce food and how we manage our land. His work combines the latest research approaches to explore innovative farm systems approaches for farm-level application.

Preface to "Enhancing Farm-Level Decision Making through Innovation"

Enhanced farm-level data sources and information in agricultural systems can allow farmers to make more timely and informed interventions that ultimately help productivity. More sustainable production practices are important for future food supplies. This Special Issue explores the use of applications that implement modelling approaches for different animal and plant systems.

Models of biological systems can be used to explore changes in climatic conditions and inform plant management options [1]. In this Special Issue, it was notable that five of the six papers investigated image analysis approaches as a means to monitor animals or plants. Image analysis has gained in interest due to continued developments in image capture technology and processing. The work of Yang Wu and Lihong Xu [2] found that image generation can improve disease detection in tomato leaves. Furthermore, Cavendish et al. [3], Waters et al. [4] and McDonagh et al. [5] showed how camera surveillance technology can be used for the high-level detection of different behaviors in cattle and sheep, with the potential to enhance management at critical periods of productive life, such as parturition. Compared to other sensor technologies, camera surveillance allows both the mother and its offspring to be monitored remotely. Shyan Ghajar and Benjamin Tracy [6] discussed how developments in proximal sensing techniques now provide opportunities to collect detailed grassland data that were previously lacking. The authors highlight that proximal sensing technologies, such as handheld sensors or sensors mounted on unmanned aerial vehicles, can provide a range of measures, such as plant species, height, biomass and nutritional content. However, the cost and complexity of new hardware and software solutions can often be barriers and hinder current adoption.

This collection of papers highlights how innovation in farming systems can support the sustainable development of food production

Funding: This research received no external funding.

Conflicts of Interest: The authors declare no conflicts of interest.

References

1. Yang, J.; He, Y.; Luo, S.; Ma, X.; Li, Z.; Lin, Z.; Zhang, Z. Optimizing the Optimal Planting Period for Potato Based on Different Water-Temperature Year Types in the Agro-Pastoral Ecotone of North China. *Agriculture* **2021**, *11*, 1061. https://doi.org/10.3390/agriculture11111061
2. Wu, Y.; Xu, L. Image Generation of Tomato Leaf Disease Identification Based on Adversarial-VAE. *Agriculture* **2021**, *11*, 981. https://doi.org/10.3390/agriculture11100981
3. Cavendish, B.; McDonagh, J.; Tzimiropoulos, G.; Slinger, K.R.; Huggett, Z.J.; Bell, M.J. Changes in Dairy Cow Behavior with and without Assistance at Calving. *Agriculture* **2021**, *11*, 722. https://doi.org/10.3390/agriculture11080722
4. Waters, B.E.; McDonagh, J.; Tzimiropoulos, G.; Slinger, K.R.; Huggett, Z.J.; Bell, M.J. Changes in Sheep Behavior before Lambing. *Agriculture* **2021**, *11*, 715. https://doi.org/10.3390/agriculture11080715
5. McDonagh, J.; Tzimiropoulos, G.; Slinger, K.R.; Huggett, Z.J.; Down, P.M.; Bell, M.J. Detecting Dairy Cow Behavior Using Vision Technology. *Agriculture* **2021**, *11*, 675. https://doi.org/10.3390/agriculture11070675

6. Ghajar, S.; Tracy, B. Proximal Sensing in Grasslands and Pastures. *Agriculture* **2021**, *11*, 740. https://doi.org/10.3390/agriculture11080740

Matt J. Bell
Editor

Communication

Detecting Dairy Cow Behavior Using Vision Technology

John McDonagh [1,*], **Georgios Tzimiropoulos** [2], **Kimberley R. Slinger** [3], **Zoë J. Huggett** [3], **Peter M. Down** [4] **and Matt J. Bell** [5]

[1] Jubilee Campus, School of Computer Science, University of Nottingham, Nottingham NG8 1BB, UK

[2] School of Electrical Engineering and Computer Science, Queen Mary University of London, London E1 4NS, UK; g.tzimiropoulos@qmul.ac.uk

[3] Sutton Bonington Campus, School of Biosciences, University of Nottingham, Sutton Bonington LE12 5RD, UK; kimberley.slinger@nottingham.ac.uk (K.R.S.); Zoe.Huggett3@nottingham.ac.uk (Z.J.H.)

[4] Sutton Bonington Campus, School of Veterinary Medicine and Science, University of Nottingham, Sutton Bonington LE12 5RD, UK; peter.down@nottingham.ac.uk

[5] Agriculture Department, Hartpury University, Gloucester GL19 3BE, UK; matt.bell@hartpury.ac.uk

* Correspondence: john.mcdonagh@nottingham.ac.uk

Abstract: The aim of this study was to investigate using existing image recognition techniques to predict the behavior of dairy cows. A total of 46 individual dairy cows were monitored continuously under 24 h video surveillance prior to calving. The video was annotated for the behaviors of standing, lying, walking, shuffling, eating, drinking and contractions for each cow from 10 h prior to calving. A total of 19,191 behavior records were obtained and a non-local neural network was trained and validated on video clips of each behavior. This study showed that the non-local network used correctly classified the seven behaviors 80% or more of the time in the validated dataset. In particular, the detection of birth contractions was correctly predicted 83% of the time, which in itself can be an early warning calving alert, as all cows start contractions several hours prior to giving birth. This approach to behavior recognition using video cameras can assist livestock management.

Keywords: dairy cows; computer vision; behaviors; monitoring; management

Citation: McDonagh, J.; Tzimiropoulos, G.; Slinger, K.R.; Huggett, Z.J.; Down, P.M.; Bell, M.J. Detecting Dairy Cow Behavior Using Vision Technology. *Agriculture* **2021**, *11*, 675. https://doi.org/10.3390/agriculture11070675

Academic Editor: Eva Voslarova

Received: 11 June 2021
Accepted: 15 July 2021
Published: 17 July 2021

Publisher's Note: MDPI stays neutral with regard to jurisdictional claims in published maps and institutional affiliations.

1. Introduction

At a time when the general public has concerns about how livestock are managed and their welfare, tools that can improve animal welfare standards and increase the public acceptance of farming are required. In recent years, the expectation has been for each stockperson to look after more animals, as input costs (including labor) have increased and finding skilled farm workers has become more challenging, and with the increased size of the average dairy herd. With these challenges have come high-quality digital camera systems that provide 24 h video surveillance capabilities, and the opportunity for farmers to monitor their livestock remotely and whilst carrying out other farm tasks. The use of cameras to monitor animals and their behaviors manually has been available for decades, with animal behavior and welfare concerns commonly directed at housed livestock production, such as dairy cows [1,2]. The monitoring of animals is essential for their welfare and survival [3].

Automated image analysis techniques have developed that allow continuous monitoring during the day and night, and require no prior training by the user other than interpreting the output. Such continuous monitoring is not possible for a stockperson. Recent technological advances in the field of computer vision based on the technique of deep learning [4,5] have emerged which now make automated monitoring of video feeds feasible. Computer vision combined with artificial intelligence (neural networks) can be used for a number of animal monitoring tasks such as recognizing the type of animals (recognition), detecting where the animals (and any other objects of interest) are located in the image (detection), localizing their body parts, and even segmenting their exact shape

1

(silhouette) from the image. Furthermore, adaptations of neural networks for analyzing video can be used for a number of tasks such as recognition of specific animal behaviors (e.g., standing, lying, walking, eating, and drinking) [6]. Major benefits of image analysis are that it does not rely on human interpretation or intervention, transponder attachments or invasive equipment (e.g., boluses and collars). Furthermore, it may provide more information compared to other monitoring systems at a relatively low cost. However, the technology does rely on obtaining a large number of high-quality images. The need for high-quality image datasets for agricultural solutions has been recognized by others [7]. Vision-based monitoring can not only detect and track individuals but also groups of animals (i.e., herd, flock or mother with offspring). Vision technology that can continuously monitor individual animals can potentially provide an objective assessment of an abnormal behavioral state to allow early intervention and improved awareness by a stockperson.

The objective of this study was to investigate using existing image recognition techniques to predict the behavior of dairy cows. This study collected a large number of high-quality video images for a range of cow behaviors. Such a dataset was found to be lacking but was required in the current study to train a computer vision model.

2. Materials and Methods

Approval for this study was obtained from the University of Nottingham animal ethics committee before commencement (approval number 151, 2017).

2.1. Data

Video cameras (5 Mp, 30 m IR. Hikvision HD Bullet; Hangzhou, China) were used to record Holstein–Friesian dairy cows at the Nottingham University Dairy Centre (Sutton Bonington, Leicestershire, UK) prior to calving. Cameras were recording at 20 frames per second, with a frame width of 640 pixels and height of 360 pixels. Three calving pens with two surveillance cameras looking into each pen were used to obtain 24 h video footage of 46 individual cows between April and June 2018. Both cameras on each pen allowed full coverage of the area (10 m × 7 m) and were approximately at a 45-degree angle looking into the pen at a height of 4 m. Each calving pen holds a maximum of eight cows. Several days prior to calving, each cow was moved into one of the three calving pens so that the entire calving process could be monitored.

2.2. Image Annotation

The video recording for each cow was annotated from 10 h before calving by three observers using custom-made scripts in the PyTorch 1.5 framework to label video clips. The PyTorch framework was used as it allows several steps in the processing of images to be carried out, such as behavioral annotations, video segmentation and model development using the Python programming language as discussed below. The start of the observation period was determined as 10 h from when the calf was fully expelled at birth using the video recording. Seven behaviors were recorded (Table 1).

Table 1. Studied behaviors and their description.

Behavior	Description
Stand	The cow is still on all four legs
Lie	The midway transition of when the cow is about to lie down to when it starts to rise again
Walk	Movement of more than two steps
Shuffle	Cow circles on the spot or moves slightly with a step or two
Contractions	Visible straining while lying down
Eating	Cow puts its head through the feeding barrier until the moment it pulls its head back out from the feeding barrier
Drinking	Head is over the water trough and regular head movement towards the trough

A total of 19,191 individual behavioral observations were obtained from all 46 cows. For the analysis, 15 video clips of each behavior that ranged between three to ten seconds were extracted from individual cow footage to provide a total of 3969 video clips for analysis (Table 2). If there were more than 15 video clips, then they would be evenly sampled from available data. There were 248–686 video clips for each behavior for training and validation. To ensure accuracy of video annotation and subsequent behavioral video clips extracted, each behavioral video clip was checked by a single trained observer to be correctly labelled and any errors corrected if required.

Table 2. Number of video clips for each behavior class in the training and validation datasets.

Label	Behavior	Training	Validation	Total
1	Stand	552	134	686
2	Lie	522	135	657
3	Walk	496	134	630
4	Shuffle	518	134	652
5	Contractions	501	112	613
6	Eating	392	91	483
7	Drinking	205	43	248
	Totals	3186	783	3969

The output of the behavior annotations from each video clip was described in a $N*3$ matrix, where N is the total number of behaviors in the video (Table 3). Start and end frames for annotated behaviors are recorded for each video clip. Each of the retained video clips were cropped to remove excessive background and to focus on a single cow (Figure 1).

Table 3. Example matrix of behavior annotations.

Start Frame	End Frame	Behavior Label
553602	556724	7
556725	557555	2
557556	557697	4
557698	580880	1
580881	581004	4
581005	581077	1
581078	581157	4

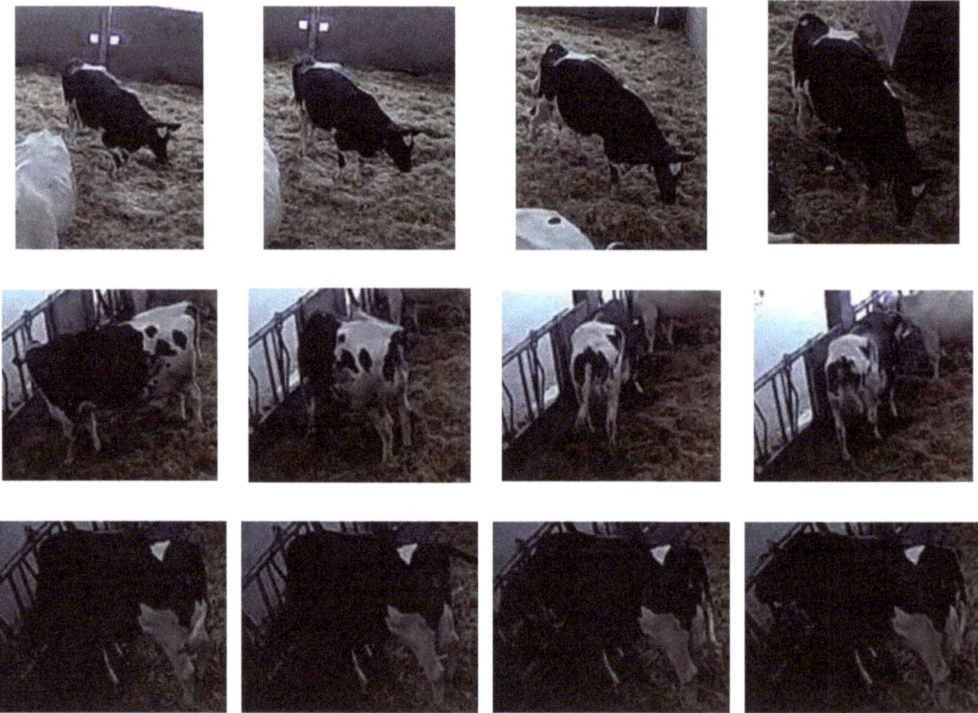

Figure 1. Example of cropped and scaled videos. Top row shows a cow walking, middle row shows a cow shuffling and bottom row is of a cow eating.

To be compliant with the non-local network [8], we used a fixed-size bounding box that fully covered the cow over all frames (this is to emulate [8], who used the entire frame). We used the image annotation tool ViTBAT [9] to generate the bounding boxes. The steps taken to process images for model development are illustrated in Figure 2.

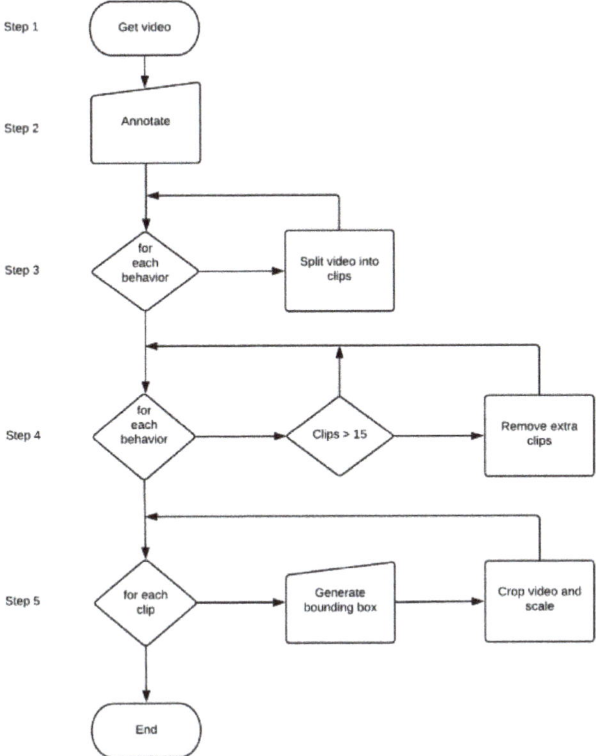

Figure 2. Illustration of steps in data acquisition and image processing.

2.3. Computer Vision Model Used for Behavior Recognition

A custom-made script in the PyTorch 1.5 framework was used to combine the behavioral matrix with cropped video images. This was performed using a non-local network [8] using the ResNet-50 architecture [10]. Further detailed explanations are discussed in prior research [8,10]. As shown in Equation (1), the non-local block computes the response at a position as a weighted sum of the features at all positions in the input feature maps and is defined as follows:

$$y_i = \frac{1}{C(x)} \sum_{\forall j} f(x_i, x_j) g(x_j), \tag{1}$$

where x is the input features, y is the output features (same size as x), i is the current position of interest, j enumerates over all possible positions, $C(x)$ is the normalization factor $C(x) = \sum_{\forall j} f(x_i, x_j)$, g is a linear embedding $g(x_j) = W_g x_j$, where W_g is learned weight matrix and $f(x_i, x_j)$ is a pairwise function that computes the correlations between the feature at location i and those at all possible positions j.

The non-local network [8] is initialized using weights that are pre-trained on the Kinetics image dataset [11], which includes 400 behaviors for humans. This approach has been shown by [12] to improve action recognition accuracy by using a pre-trained initialization starting point for modelling. To decrease training and testing times, the current study used 8-frame input clips. The 8-frame clips were generated by randomly cropping out 64 consecutive frames from the training video and then keeping 8 frames that are evenly separated by a stride of 8 frames (Figure 3). Additionally, while training, the spatial size is fixed to 224 pixels squared, which is randomly cropped from a video or its horizontal flip, whose shorter side is randomly scaled between 256 and 320 pixels.

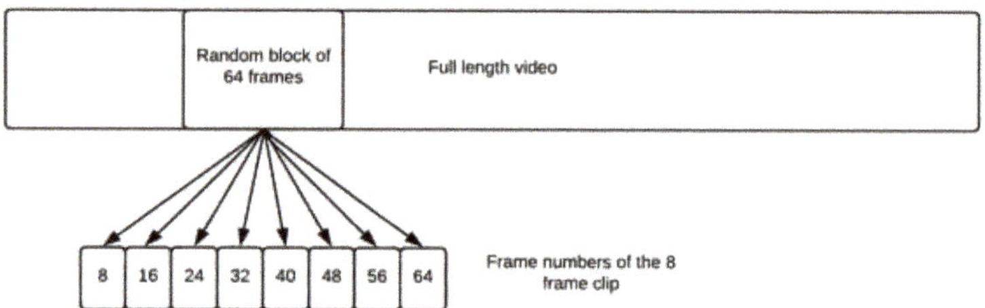

Figure 3. Temporal sampling of each video clip with eight evenly spaced frames being selected from a block of 64 consecutive frames.

2.4. Computer Vision Model Validation

To validate the performance of the model, we performed spatially fully convolutional inference as described by [8]. Briefly, the shorter side is resized to 256 pixels and 3 crops of 256x256 pixels are used to cover the entire spatial domain. The final predicted output is the average score for 10 evenly spaced 8 frame clips sampled along the temporal dimension of a full-length video (Figure 4).

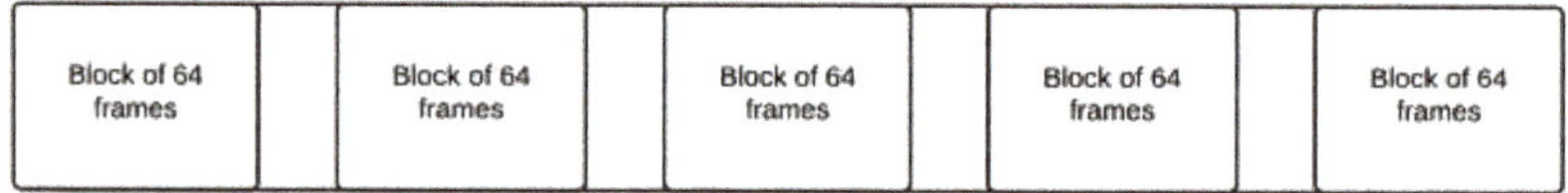

Figure 4. Ten clips of eight frames are sampled from blocks (64 frames) which are evenly sampled over the entire video. Each clip produces its own score, and the final output is the average of all the scores (a total of 5 blocks are shown for illustration purposes.)

3. Results and Discussion

Despite scientific value, pressing need and direct impact on animal health and welfare, very little attention has been paid in developing an annotated video dataset of dairy cow behaviors. Most research to date has been based on wearable accelerometer-based activity monitoring sensors [13–15]. We introduce a new large-scale video dataset for the purpose of cow behavior classification. Image banks containing a large number of high-quality (i.e., accurate and high-resolution) images for different applications are needed to develop vision-based technologies, such as behavior recognition in animals, as suggested by other studies [7]. This study showed that automated monitoring of the cow during parturition is possible, which for a high-value animal is beneficial to assist the stockperson and enhance animal welfare.

Our dataset consisted of almost 4000 video clips of individual animal behaviors, each between 3 and 10 s in length, which were on pregnant dairy cows prior to calving. There was over 9 h and 42 min of captured video data, which was split into approximately 7 h and 48 min for training and 1 h and 54 min for validation. In the field of computer vision, action recognition has been applied on humans with a high degree of success [8]. We show that the same model pre-trained on a dataset devised for human action recognition, namely Kinetics [11], can be successfully adapted to detect the behavior of dairy cows. As shown in Table 4, the accuracy of identifying contractions while lying was 83%—this in itself is sufficient enough to predict the birth of a calf, as a cow will generally start contractions approximately 1 to 2 h prior to giving birth. Standing, lying, eating and drinking behaviors all scored greater that 84% and can also help with the monitoring of animal well-being. Furthermore, changes in duration or frequency of behaviors studied may help identify abnormal behavior patterns that can assist in animal management. For example, eating and drinking can be detected with a high level of accuracy at over 90%, and these behaviors can be used to identify health problems [16].

Table 4. Evaluation of model predictions against validation dataset.

		Stand	Lie	Walk	Shuffle	Contractions	Eating	Drinking
Target [1]	No.	134	135	134	134	112	91	43
Output [2]	No.	113	122	107	108	93	86	40
Accuracy [3]	%	84	90	80	80	83	95	93

[1] The target row shows how many video clips were tested for each behavior. [2] Output row shows how many behavior video clips the model classified correctly. [3] The percentage of target behavior video clips correctly classified.

As well as working with cows, the proposed computer vision approach could be adapted for other livestock species such as pigs, poultry, sheep, and horses to predict birth and identify behavior patterns or behaviors that occur over many hours, which may be missed by subjective and observational sampling. Furthermore, because the calving pen is continuously monitored, it should also be possible to detect and track the behaviors of the mother and its newborn offspring, which is not feasible using standard predictive animal monitoring applications that are currently being used by the livestock industry.

The development of behavior recognition using continuous camera surveillance within the farm environment is challenging. The current study identified several potential causes of error in computer model predictions which are limitations of current vision-based monitoring (Table 5).

Table 5. Potential causes of error in animal vision-based model predictions.

Problem	Cause of Error
Pose	A cow's pose changes not only in terms of its current behavior, but also in terms of the direction it is facing from the camera. As a cow is a quadruped, this forces the model to have a much higher generalization capability when compared to bipeds such as humans.
Similarity	Distinguishing between two or more cows is a very difficult task even for humans. This is because cows can often have similar colors or patch patterns on their bodies.
Occlusion	Parts of a cow can be hidden if behind other cows, such as when all bunched up while eating. The birth of the calf can also be occluded if the cow is facing towards the surveillance camera. Cows can also be partially hidden under bedding. Cows can even have self-occlusion, where the cow's body blocks the view to other parts such as the head. Spider webs can also blur/occlude cows while the camera is in infrared night vision mode.
Lighting	Natural light comes through the ventilation spaces, which can produce rectangular patches over the enclosure and on the cows. Over the course of the day, the brightness of the enclosure changes. In the evening artificial lighting is used, which gives an orange tint to the enclosure. Infrared night vision is used during night-time, which turns the video footage into black and white. While the camera is in night vision mode, it focuses on the center of the pen and loses focus towards the extremities of the enclosure. Night vision also casts deep shadows off the cows that may confuse object detection.

4. Conclusions

We show that computer vision can be successfully applied to predict individual dairy cow behaviors with an accuracy of 80% or more for the behaviors studied. This approach could be used for early detection of abnormal behavior in animals, birth events and the need for assistance. Computer vision technology may help a stockperson make more timely decisions based on the continuous tracking of individuals within groups of animals.

Author Contributions: Conceptualization, M.J.B. and G.T.; methodology, M.J.B. and G.T.; software, J.M.; validation, J.M.; formal analysis, J.M.; investigation, J.M.; resources, M.J.B.; data curation, M.J.B., K.R.S., Z.J.H. and J.M.; writing—original draft preparation, J.M. and M.J.B.; writing—review and editing, J.M., M.J.B., G.T. and P.M.D.; visualization, J.M.; supervision, M.J.B. and G.T.; project administration, M.J.B.; funding acquisition, M.J.B. and G.T. All authors have read and agreed to the published version of the manuscript.

Funding: This research was funded by the Douglas Bomford Trust, the Engineering and Physical Sciences Research Council and the Biotechnology and Biological Sciences Research Council.

Institutional Review Board Statement: The study was conducted according to the guidelines of the Declaration of Helsinki, and approved by the Animal Ethics Committee at the University of Nottingham (approval number 151, 2017).

Informed Consent Statement: Not applicable.

Data Availability Statement: The analyzed datasets are available from the corresponding author on request.

Conflicts of Interest: The authors declare no conflict of interest.

References

1. Lawrence, A.; Stott, A. Profiting from animal welfare: An animal-based perspective. *J. R. Agric. Soc. Engl.* **2009**, *170*, 40–47.
2. Barkema, H.W.; Von Keyserlingk, M.A.G.; Kastelic, J.P.; Lam, T.J.G.M.; Luby, C.; ROY, J.P.; Leblanc, S.J.; Keefe, G.P.; Kelton, D.F. Invited review: Changes in the dairy industry affecting dairy cattle health and welfare. *J. Dairy Sci.* **2015**, *98*, 7426–7445. [CrossRef] [PubMed]
3. Wathes, C.M.; Kristensen, H.H.; Aerts, J.-M.; Berckmans, D. Is precision livestock farming an engineer's daydream or nightmare, an animal's friend or foe, and a farmer's panacea or pitfall? *Comput. Electron. Agric.* **2008**, *64*, 2–10. [CrossRef]
4. Krizhevsky, A.; Sutskever, I.; Hinton, G. Imagenet classification with deep convolutional neural networks. *Adv. Neural Inf. Process. Syst.* **2012**, *25*, 1097–1105. [CrossRef]
5. Girshick, R.; Donahue, J.; Darrell, T.; Malik, J. Rich feature hierarchies for accurate object detection and semantic segmentation. In Proceedings of the IEEE Conference on Computer Vision and Pattern Recognition, Columbus, OH, USA, 24–27 June 2014.
6. Cangar, Ö.; Leroy, T.; Guarino, M.; Vranken, E.; Fallon, R.; Lenehan, J.; Mee, J.; Berckmans, D. Automatic real-time monitoring of locomotion and posture behavior of pregnant cows to calving using online image analysis. *Comput. Electron. Agric.* **2008**, *64*, 53–60. [CrossRef]
7. Tian, H.; Wang, T.; Liu, Y.; Qiao, X.; Li, Y. Computer vision technology in agricultural automation—A review. *Inf. Process. Agric.* **2020**, *7*, 1–19. [CrossRef]
8. Wang, X.; Girshick, R.; Gupta, A.; He, K. Non-Local Neural Networks. *arXiv* **2018**, arXiv:1711.07971v3.
9. Biresaw, T.; Nawaz, T.; Ferryman, J.; Dell, A. ViTBAT: Video Tracking and Behavior Annotation Tool. In Proceedings of the 13th IEEE International Conference on Advanced Video and Signal Based Surveillance (AVSS), Colorado Springs, CO, USA, 23–26 August 2016; pp. 295–301. [CrossRef]
10. He, K.; Zhang, X.; Ren, S.; Sun, J. Deep Residual Learning for Image Recognition. *arXiv* **2015**, arXiv:1512.03385v1.
11. Kay, W.; Carreira, J.; Simonyan, K.; Zhang, B.; Hillier, C.; Vi Jayanarasimhan, S.; Viola, F.; Green, T.; Back, T.; Natsev, P.; et al. The kinetics human action video dataset. *arXiv* **2017**, arXiv:1705.06950.
12. Carreira, J.; Zisserman, A. Quo Vadis, Action Recognition? A New Model and the Kinetics Dataset. *arXiv* **2017**, arXiv:1705.07750v3.
13. Diosdado, J.; Barker, Z.; Hodges, H.; Amory, J.; Croft, D.; Bell, N.; Codling, E. Classification of behavior in housed dairy cows using an accelerometer-based activity monitoring system. *Anim. Biotelem.* **2015**, *3*, 15. [CrossRef]
14. Rahman, A.; Smith, D.; Little, B.; Ingham, A.; Greenwood, P.; Bishop-Hurley, G. Cattle behavior classification from collar, halter, and ear tag sensors. *Inf. Process. Agric.* **2018**, *5*, 124–133.
15. Benaissa, S.; Tuyttens, F.; Plets, D.; Pessemier, T.; Trogh, J.; Tanghe, E.; Martens, L.; Vandaele, L.; van Nuffel, A.; Joseph, W.; et al. On the use of on-cow accelerometers for the classification of behaviors in dairy barns. *Res. Vet. Sci.* **2019**, *125*, 425–433. [CrossRef] [PubMed]
16. Weary, D.M.; Huzzey, J.M.; von Keyserlingk, M.A.G. Board-Invited review: Using behavior to predict and identify ill health in animals. *J. Anim. Sci.* **2009**, *87*, 770–777. [CrossRef] [PubMed]

Article

Changes in Sheep Behavior before Lambing

Beatrice E. Waters [1], John McDonagh [2], Georgios Tzimiropoulos [3], Kimberley R. Slinger [1], Zoë J. Huggett [1] and Matt J. Bell [4,*]

[1] School of Biosciences, University of Nottingham, Sutton Bonington Campus, Sutton Bonington LE12 5RD, UK; Beatrice.Waters@nottingham.ac.uk (B.E.W.); kimberley.slinger@nottingham.ac.uk (K.R.S.); Zoe.Huggett3@nottingham.ac.uk (Z.J.H.)
[2] School of Computer Science, University of Nottingham, Jubilee Campus, Nottingham NG8 1BB, UK; john.mcdonagh@nottingham.ac.uk
[3] School of Electrical Engineering and Computer Science, Queen Mary University of London, London E1 4NS, UK; g.tzimiropoulos@qmul.ac.uk
[4] Agriculture Department, Hartpury University, Gloucester GL19 3BE, UK
* Correspondence: matt.bell@hartpury.ac.uk

Abstract: The aim of this study was to assess the duration and frequency of behavioral observations of pregnant ewes as they approached lambing. An understanding of behavioral changes before birth may provide opportunities for enhanced visual monitoring at this critical stage in the animal's life. Behavioral observations for 17 ewes in late pregnancy were recorded during two separate time periods, which were 4 to 6 weeks before lambing and before giving birth. It was normal farm procedure for the sheep to come indoors for 6 weeks of close monitoring before lambing. The behaviors of standing, lying, walking, shuffling and contraction behaviors were recorded for each animal during both time periods. Over both time periods, the ewes spent a large proportion of their time either lying (0.40) or standing (0.42), with a higher frequency of standing (0.40) and shuffling (0.28) bouts than other behaviors. In the time period before giving birth, the frequency of lying and contraction bouts increased and the standing and walking bouts decreased, with a higher frequency of walking bouts in ewes that had an assisted lambing. The monitoring of behavioral patterns, such as lying and contractions, could be used as an alert to the progress of parturition.

Keywords: behavior; birth; management; observations; sheep

Citation: Waters, B.E.; McDonagh, J.; Tzimiropoulos, G.; Slinger, K.R.; Huggett, Z.J.; Bell, M.J. Changes in Sheep Behavior before Lambing. *Agriculture* **2021**, *11*, 715. https://doi.org/10.3390/agriculture11080715

Academic Editors: Claudia Arcidiacono and Eva Voslarova

Received: 17 June 2021
Accepted: 27 July 2021
Published: 29 July 2021

Publisher's Note: MDPI stays neutral with regard to jurisdictional claims in published maps and institutional affiliations.

1. Introduction

A stockperson's ability to assess animal behavior is a key component of their ability to recognize and treat ill-health, and evaluate the wellbeing of their livestock [1,2]. The visual assessment of livestock by humans is subjective and has several limitations such as the cost of labor and time to regularly observe individual animals. Hence, several monitoring technologies have been proposed in recent years that predict animal behaviors from movement sensors on cattle or sheep [3–8]. New technologies that provide an objective measure of animal behavior, such as sensors and cameras, could provide an aid to improve animal management [9]. Furthermore, monitoring equipment can continuously and remotely track livestock, something that would be unrealistic and too costly for human observers to replicate [3].

Lambing is a critical time in the productive life of sheep and the development of the newborn offspring that will eventually be sold or retained as flock replacements. Sheep will often have multiples at birth, which can be physically challenging, stressful and a painful process for the mother and offspring that may require a farmer's intervention [10]. The parturition period is associated with several physiological, hormonal and behavioral changes in the pregnant animal, with restless behavior exhibited by nesting and reduced appetite along with birth contractions, which increase in frequency and intensity as birth progresses [10,11]. Studying cows, Huzzy et al. [12] found a dramatic increase in the number of positional changes such as lying or standing at calving, and reported that the animals tended to isolate themselves from the rest of the herd. Although there are several studies on changes in cattle behavior before calving [10,12–15], there are few studies on pregnant sheep [16,17]. The need for further research into enhanced monitoring approaches of sheep during parturition has been identified by others [17], and assist a stockperson during this critical period. The hypothesis of the current study was that there is a change in sheep behavior before giving birth, and this change can be visually observed. This insight may assist lambing management and future monitoring technologies.

The objective of this study was to assess the duration and frequency of behavioral observations of pregnant sheep as they approached lambing. The sheep studied followed normal husbandry procedure of being housed as a group at about 6 weeks before lambing to allow for closer monitoring.

2. Materials and Methods

Approval for this study was obtained from the University of Nottingham animal ethics committee before the study commenced (approval number 198, 2018).

2.1. Data

A total of 17 pregnant ewes were monitored using video camera surveillance (5 Mp, 30 m IR. Hikvision HD Bullet; Hangzhou, China) at the Nottingham University Farm (Sutton Bonington, Leicestershire, UK; 52.8282° N 1.2485° W, 48 m a.s.l) when indoors before lambing from February to March 2019. The study was designed to have similar numbers of primiparous and multiparous ewes. The ewes monitored were predominantly Lleyn breed, with 9 primiparous and 8 multiparous. Normal husbandry procedure for the flock were followed, whereby all sheep came indoors for closer monitoring as a single group at about 6 weeks before lambing, and returned to pasture after lambing. The sheep were group housed on straw bedding, with an open feed trough for forage and supplementary feed and a single water trough. A single camera was used to obtain continuous video footage of each ewe. The camera position allowed full coverage of the area and at an approximate 45-degree angle looking into the sheep pen. When the sheep were housed at the start of the study they were weighed, marked with a number for camera observations and individual identification recorded, and vaccinated for pasteurella and clostridial disease, but after this the sheep were not handled until they had given birth. The sheep did not receive any other health treatments, such for endo or ectoparasites, during the study. The average age of primiparous ewes was 1.9 (s.d. 0.03) years and multiparous 4.3 (s.d. 1.0) years. The average bodyweight of primiparous ewes was 57.9 (s.d. 2.6) kg and multiparous 64.2 (s.d. 6.4) kg. Sheep were group fed *ad libitum* haylage consisting of 9.7 MJ/kg for metabolizable energy (ME), 486, 548, 138, 59 g/kg for dry matter, neutral detergent fiber, crude protein and sugar, respectively (Sciantec Analytical Services, Cawood, UK; using near-infrared spectroscopy analysis). Additionally, sheep were supplemented with 350 g/day oats with wheat distillers grain mix (13.4 MJ/kg ME, 860, 250, 228 and 60 g/kg for dry matter, neutral detergent fiber, crude protein and sugar, respectively). The diet was about 75% forage on a dry matter basis. The feed was allocated as 2% of the average bodyweight for the group of ewes (about 1.2 kg/day), with the diet formulated based on an estimated energy and protein requirement of 11 MJ/kg ME and 160 g/kg crude protein in the diet [18]. The same diet was fed throughout the study and the amount allocated to the group was reduced as sheep gave birth and were removed from the lambing pen. Need for a birth to be assisted by farm staff was recorded for each ewe. The average daily temperature was 5.7 °C, rainfall was 2.9 mm and humidity was 90% during the study.

2.2. Observations

Two observation periods were used to investigate changes in behavior: Period 1 (4–6 weeks before lambing) and Period 2 (at lambing). There were 10 h of annotated video recordings for each ewe from Period 1 and three hours before the first lamb was born for Period 2. Tracking of sheep in Period 2 before lambing was challenging because of the similar appearance and behavioral changes. Three observers used custom made scripts in PyTorch 1.5 framework to record the behavior profile of each ewe with time. A total of 8257 individual behavioral observations were recorded from all 17 ewes. To ensure accuracy of video behavior annotations, the video was segmented into short clips for each behavior, and all video clips subsequently checked for accuracy by one of the three observers. Five behaviors were recorded, which were:

1. Standing: The sheep is still on all four legs.
2. Lying: The midway transition of when the sheep is about to lie down to when they start to rise again.
3. Walking: Movement of more than two steps.
4. Shuffle: Sheep circles on the spot or moves slightly with a step or two.
5. Contractions: Visible straining while lying down.

2.3. Statistical Analysis

The duration of behaviors in seconds and behavior frequency were determined for both time periods. A total of 170 behavior records were obtained from 17 ewes (17 × 5 behaviors × 2 time periods).

Behavior records were analyzed using a generalized linear mixed model in Genstat Version 19.1 (Lawes Agricultural Trust, 2018). A binomial error distribution and a logit link function was fitted to the fixed effects of assistance, time period, behavior and parity for the dependent variables of duration and frequency of behaviors in Equation (1):

$$Y_{ijkl} = \mu + A_i \times T_j \times B_k + P_l + E_{ijkl} \tag{1}$$

where Y_{ijkl} is the dependent variable of behavior duration or frequency; μ = overall mean; A_i = fixed effect for assistance at lambing (i = 0 for unassisted or 1 for assisted); T_j = fixed effect of time period (j = 1 or 2); B_k = fixed effect of behavior (k = standing, lying, walking, shuffling, contractions); P_l = fixed effect of parity (l = primiparous or multiparous); E_{ijkl} = random error term.

The back-transformed predicted means for behavior duration and frequency were expressed as the proportion of total time or count during each time period. Significance was attributed at $p < 0.05$.

3. Results

Of the 17 lambings, two were triplets, 13 were twins and two were singles. There were four primiparous and two multiparous ewes that required assistance by the farm stockperson, with all other lambings being unassisted.

Differences were found in the duration of behaviors ($p < 0.001$) with most of the time spent either lying (0.40) or standing (0.42), with other behaviors being 0.08 or less across time periods studied (Figure 1).

There was no effect of parity, time period, lambing assistance on duration of behaviors ($p > 0.05$; Table 1 and Figure 2).

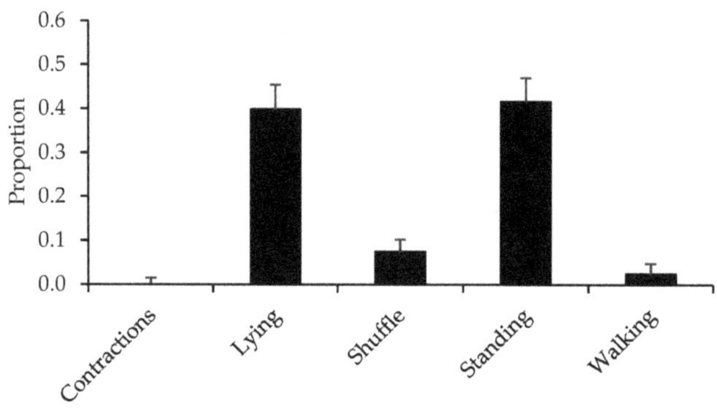

Figure 1. Predicted mean (±SEM) proportion of time spent in doing different behaviors.

Table 1. Effects of parity, time period, and lambing assistance on the duration of behaviors as a proportion of time.

Variable		Mean (s.e.)	Df	F Statistic	*p* Value [5]
Parity	Primiparous	0.01 (77)	1	0.0000002	0.999
	Multiparous	0.01 (77)			
Time period [1]	Period 1	0.001 (14.3)	1	0.0000002	0.999
	Period 2	0.13 (0.05)			
Assistance [2]	Assisted	0.01 (78.5)	1	0.002	0.962
	Unassisted	0.01 (77.2)			
Behavior [3]			4	8.7	<0.001
Time period × assistance	Period 1/Assisted	0.001 (13.1)	1	0.04	0.850
	Period 1/Unassisted	0.001 (15.6)			
	Period 2/Assisted	0.14 (0.09)			
	Period 2/Unassisted	0.13 (0.05)			
Assistance × behavior	Assisted/Contractions	0.000001 (0.01)	4	0.04	0.998
	Assisted/Lying	0.41 (0.08)			
	Assisted/Shuffle	0.07 (0.04)			
	Assisted/Standing	0.43 (0.07)			
	Assisted/Walking	0.02 (0.02)			
	Unassisted/Contractions	0.000001 (0.02)			
	Unassisted/Lying	0.38 (0.08)			
	Unassisted/Shuffle	0.08 (0.04)			
	Unassisted/Standing	0.40 (0.08)			
	Unassisted/Walking	0.03 (0.05)			
Time period × behavior	Period 1/Contractions	0 (0)	4	0.5	0.715
	Period 1/Lying	0.48 (0.04)			
	Period 1/Shuffle	0.07 (0.02)			
	Period 1/Standing	0.40 (0.04)			
	Period 1/Walking	0.04 (0.02)			
	Period 2/Contractions	0.14 (0.07)			
	Period 2/Lying	0.32 (0.09)			
	Period 2/Shuffle	0.09 (0.06)			
	Period 2/Stand	0.43 (0.10)			
	Period 2/Walking	0.02 (0.03)			
Time period × behavior × assistance [4]			3	0.5	0.685

[1] Period 1 was observations obtained 4 to 6 weeks before lambing when the sheep came indoors for close monitoring, and Period 2 was observations obtained before the ewe gave birth. [2] Births were either assisted or unassisted by farm staff. [3] Predicted mean values shown in Figure 1. [4] Predicted mean values shown in Figure 2. [5] Significance was attributed at *p* < 0.05.

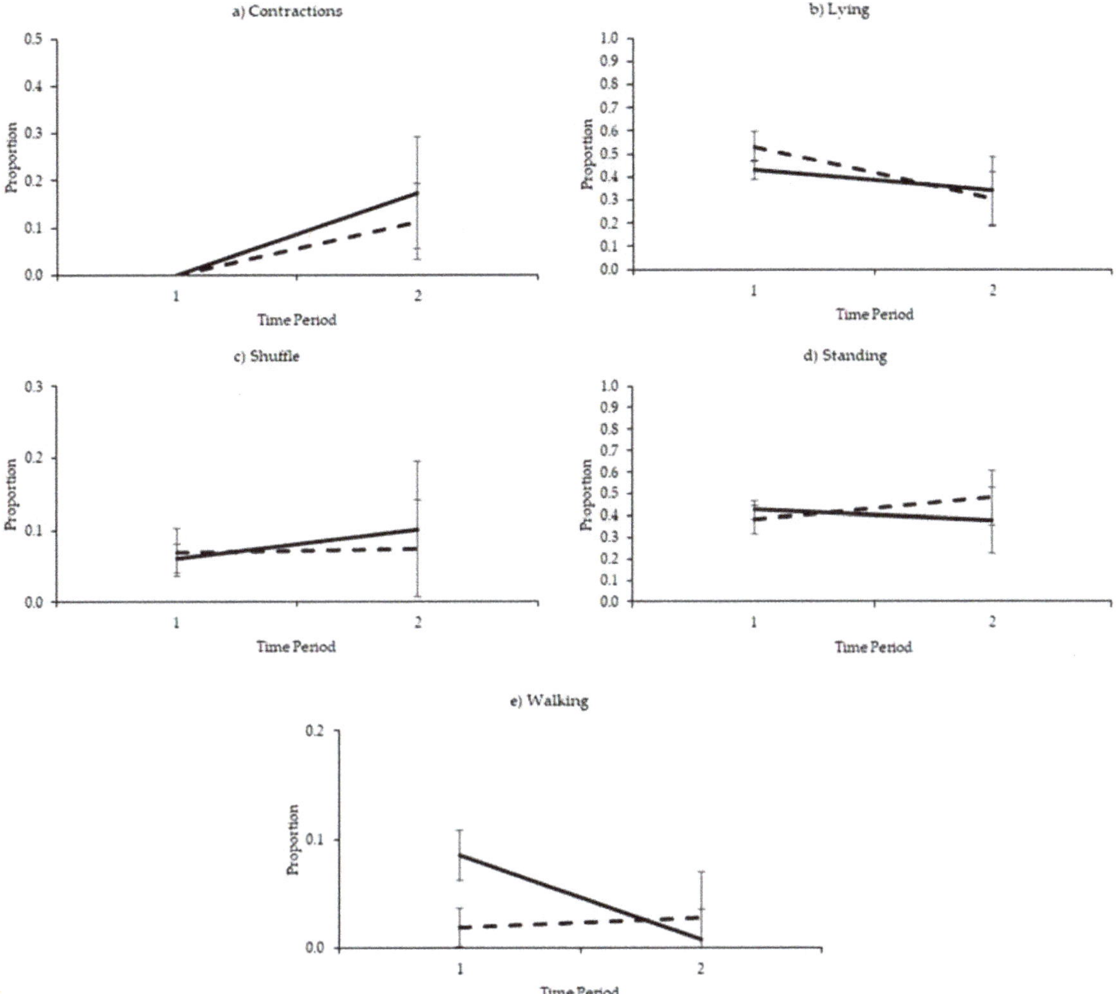

Figure 2. Predicted mean (±SEM) proportion of time that were (**a**) contractions, (**b**) lying, (**c**) shuffling, (**d**) standing and (**e**) walking behavior for assisted (dashed line) or non-assisted (solid line) lambing in time Periods 1 or 2, with Period 2 ending with the lambing event.

Differences were also found in the frequency of behaviors ($p < 0.001$) with standing (0.40) and shuffling (0.28) being the most frequent, with other behaviors being 0.09 or less across time periods studied (Figure 3).

In the time period before lambing, the frequency of lying and contraction bouts increased and the standing and walking bouts decreased ($p < 0.001$; Table 2), with a higher frequency of walking bouts in sheep that had an assisted lambing ($p < 0.01$; Figure 4). There was no effect of parity on frequency of behaviors.

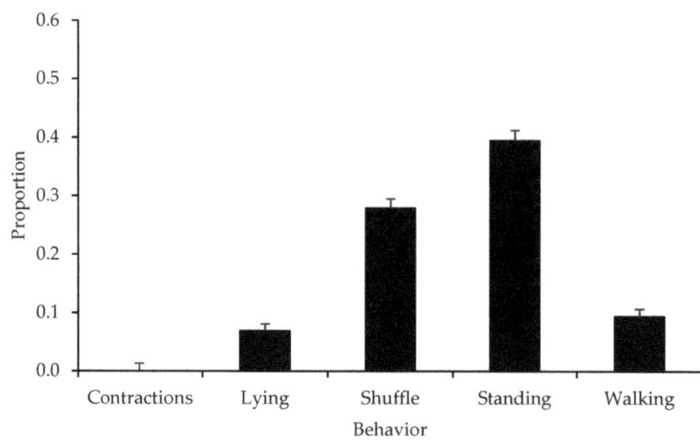

Figure 3. Predicted mean (±SEM) proportion of observations for different behaviors.

Table 2. Effects of parity, time period, and lambing assistance on the frequency of behaviors as a proportion of observations.

Variable		Mean (s.e.)	Df	F Statistic	p Value [5]
Parity	Primiparous	0.02 (41.4)	1	0.0000001	0.999
	Multiparous	0.02 (41.4)			
Time period [1]	Period 1	0.002 (7.8)	1	0.000001	0.999
	Period 2	0.17 (0.01)			
Assistance [2]	Assisted	0.02 (38.7)	1	1.5	0.219
	Unassisted	0.02 (44.3)			
Behavior [3]			4	53.4	<0.001
Time period × assistance	Period 1/Assisted	0.002 (8.3)	1	0.05	0.822
	Period 1/Unassisted	0.002 (7.4)			
	Period 2/Assisted	0.18 (0.01)			
	Period 2/Unassisted	0.16 (0.02)			
Assistance × behavior	Assisted/Contractions	0.000001 (0.01)	4	0.7	0.612
	Assisted/Lying	0.06 (0.02)			
	Assisted/Shuffle	0.29 (0.03)			
	Assisted/Standing	0.41 (0.03)			
	Assisted/Walking	0.12 (0.02)			
	Unassisted/Contractions	0.000001 (0.01)			
	Unassisted/Lying	0.08 (0.02)			
	Unassisted/Shuffle	0.27 (0.02)			
	Unassisted/Standing	0.38 (0.02)			
	Unassisted/Walking	0.08 (0.02)			
Time period × behavior	Period 1/Contractions	0 (0)	4	17.1	<0.001
	Period 1/Lying	0.03 (0.01)			
	Period 1/Shuffle	0.30 (0.03)			
	Period 1/Standing	0.46 (0.03)			
	Period 1/Walking	0.20 (0.02)			
	Period 2/Contractions	0.19 (0.02)			
	Period 2/Lying	0.17 (0.02)			
	Period 2/Shuffle	0.26 (0.02)			
	Period 2/Standing	0.33 (0.02)			
	Period 2/Walking	0.04 (0.01)			
Time period × behavior × assistance [4]			3	5.0	<0.01

[1] Period 1 was observations obtained 4 to 6 weeks before lambing when the sheep came indoors for close monitoring, and Period 2 was observations obtained before the ewe gave birth. [2] Births were either assisted or unassisted by farm staff. [3] Predicted mean values shown in Figure 3. [4] Predicted mean values shown in Figure 4. [5] Significance was attributed at $p < 0.05$.

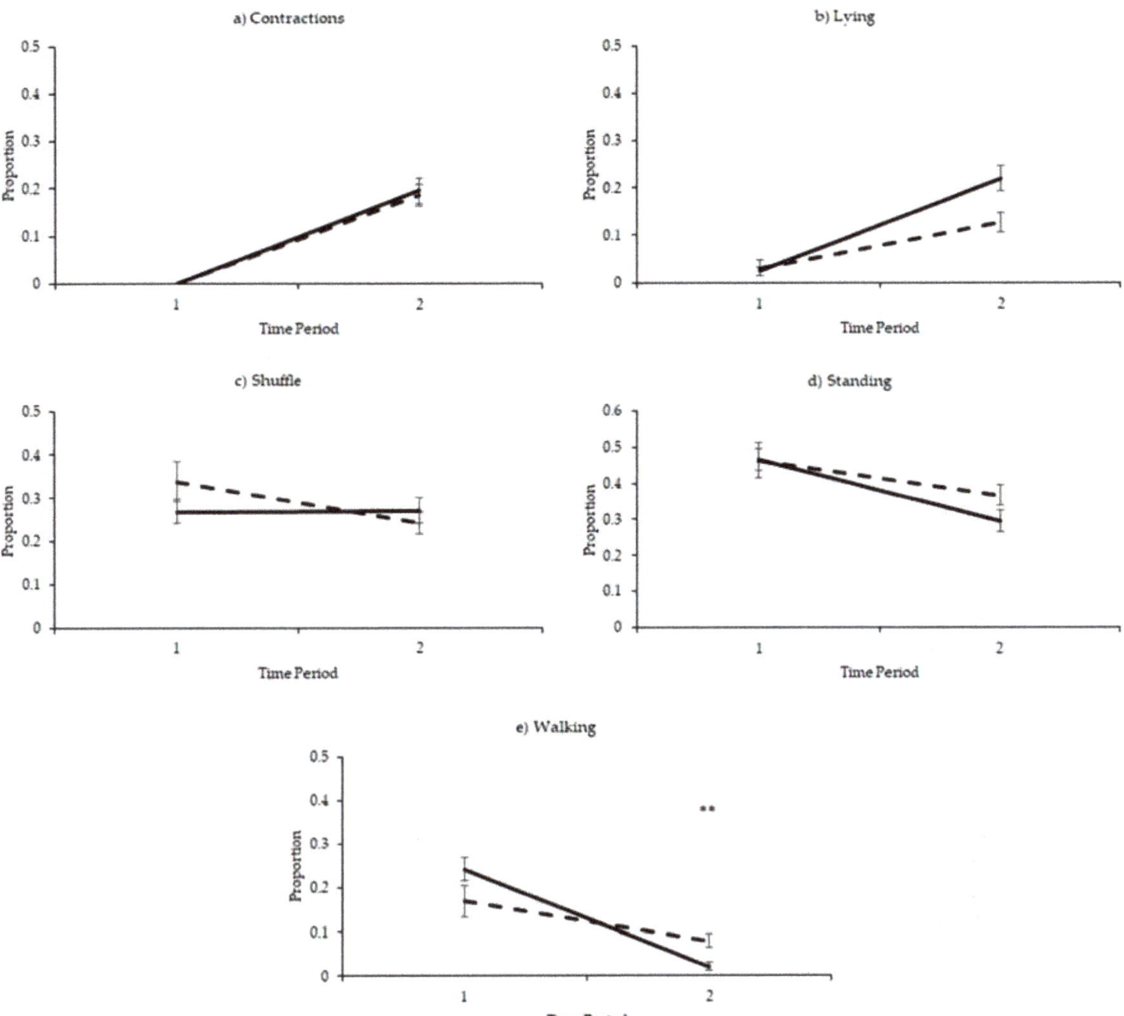

Figure 4. Predicted mean (±SEM) proportion of observations that were (**a**) contractions, (**b**) lying, (**c**) shuffling, (**d**) standing and (**e**) walking behavior for assisted (dashed line) or unassisted (solid line) lambing in time Periods 1 or 2, with Period 2 ending with the lambing event. ** $p < 0.01$.

4. Discussion

The current study found that sheep spend most of their time either standing or lying during pregnancy. There are surprisingly few studies investigating the duration and frequency of behaviors of pregnant sheep. The current study suggests pregnant ewes spend about 10 h per day lying and 10 h per day standing. This is similar to other ruminant animals such as cattle when indoors, which have been found to spend between 10–12 h per day lying [19]. The duration of behaviors did not appear to change during parturition; however, the frequency of lying bouts, including contractions, increased in the period before lambing. The sheep in the current study were within the last trimester of pregnancy, and the lack of general movement can be expected in heavily pregnant animals with little distance to their food. Previous studies have shown that there are few differences in the behavior of ewes before and during parturition because of factors such as breed, age of ewe, nutrition, climate or location [16]. The current study also found no significant effect of parity. In cows, Barrier et al. [13] found assisted animals displayed more frequent contractions than those that were unassisted, and that this was often because of awkward positioning of the calf at birth. A general state of restlessness is common in animals as parturition approaches and can also be seen in cow studies, characterized by an increase in lying frequencies as observed in the current study, general increased activity, reduced feed intake, and an intensified stress response [12,14]. In the current study, inclusion of additional behaviors, such as eating and drinking, would have been useful information as these behaviors are known to change as lambing approaches. However, they could not reliably be observed in sheep that were group housed as in the current study. A study by Black and Krawczel [20] found a higher lying frequency was associated with difficult calvings and that the cows that were not exercised were more uncomfortable during parturition. The sheep with an assisted birth in the current study had a higher frequency of walking bouts compared to unassisted births, which suggests restless behavior patterns. Generally, the sheep before lambing in the current study reduced their standing and walking bouts as lying bouts increased. Fogarty et al. [17] found a general increase in walking behavior and frequency of posture changes (i.e., standing and lying) in ewes before lambing. The change in frequencies of standing, lying and walking may provide useful indicators for tracking the progression of birth.

Sheep are often managed in large groups. This makes close inspection of individual animals difficult and timing of observations important for animal husbandry. Therefore, to enhance a farmer's capacity to manage individual animals in large groups, and detect animals who are ill or injured, animal tracking technology has been developed [21,22]. Sensor technologies are not however free from challenges; these devices are extremely sensitive and can be prone to damage from the dirt and dust that comes with farm environments. Their success will rely on the cost-benefit for livestock farming and added value to farm operations such as supporting the intense monitoring of parturition during day or night periods [23]. Sensor technologies could benefit sheep production by allowing more frequent and effective behavior observations at this key stage. Increased behavior monitoring would be extremely beneficial during parturition, as mortality will affect both animal welfare and farm productivity. In sheep, accelerometers have previously been used to detect behavioral states such as high and low general activity or some combinations of lying, standing, grazing, walking and/or running [3,17,24]. Use of an accelerometer, with machine learning, has been found to accurately predict 91% of lambing events within 3 h of birth based on body posture alone [25]. Sensors detecting key behaviors, such as those studied, present new opportunities for a continuous and real-time objective measurement in farm animals [23].

The current study involved a relatively small flock of 17 ewes because of challenges associated with complete video surveillance of animals over two time periods, and obstruction of view in group housing. Although the number of animals studied may have affected the results of this study, they appear consistent with other animal studies as mentioned above. Multiple cameras may have helped increase surveillance coverage and increased the number of animals studied. However, the results from this study suggest that observing changes in lying bouts and detection of contractions could assist farmers in monitoring parturition to enhance sheep husbandry.

5. Conclusions

This current study investigating group housed ewes during late pregnancy found an increased frequency of lying bouts, including contractions, before lambing. Pregnant ewes spent a large proportion of their time either lying or standing, with a higher frequency of standing and shuffling bouts. Ewes that required assistance at lambing had more walking bouts compared to ewes that were unassisted. The monitoring of behavioral patterns, such as lying and contractions, could be used as an alert to the progress of parturition.

Author Contributions: Conceptualization, M.J.B. and G.T.; methodology, M.J.B. and G.T.; software, J.M.; validation, J.M.; formal analysis, B.E.W. and M.J.B.; investigation, B.E.W. and M.J.B.; resources, M.J.B.; data curation, M.J.B., K.R.S., Z.J.H. and J.M.; writing—original draft preparation, B.E.W. and M.J.B.; writing—review and editing, B.E.W. and M.J.B.; visualization, J.M.; supervision, M.J.B.; project administration, M.J.B.; funding acquisition, M.J.B. and G.T. All authors have read and agreed to the published version of the manuscript.

Funding: This research was funded by the Douglas Bomford Trust, the Engineering and Physical Sciences Research Council and the Biotechnology and Biological Sciences Research Council.

Institutional Review Board Statement: The study was conducted according to the guidelines of the Declaration of Helsinki, and approved by the Animal Ethics Committee at the University of Nottingham (approval number 198, 2018).

Informed Consent Statement: Not applicable.

Data Availability Statement: The analyzed datasets are available from the corresponding author on request.

Conflicts of Interest: The authors declare no conflict of interest.

References

1. Dawkins, M.S. Behaviour as a tool in the assessment of animal welfare. *Zoology* **2003**, *106*, 383–387. [CrossRef]
2. Weary, D.M.; Huzzey, J.M.; von Keyserlingk, M.A.G. Board-invited review: Using behavior to predict and identify ill health in animals. *J. Anim. Sci.* **2009**, *87*, 770–777. [CrossRef] [PubMed]
3. Barwick, J.; Lamb, D.W.; Dobos, R.; Welch, M.; Trotter, M. Categorising sheep activity using a tri-axial accelerometer. *Comput. Electron. Agric.* **2018**, *145*, 289–297. [CrossRef]
4. Carslake, C.; Vázquez-Diosdado, J.A.; Kaler, J. Machine learning algorithms to classify and quantify multiple behaviours in dairy calves using a sensor: Moving beyond classification in precision livestock. *Sensors* **2020**, *21*, 88. [CrossRef] [PubMed]
5. Diosdado, J.; Barker, Z.; Hodges, H.; Amory, J.; Croft, D.; Bell, N.; Codling, E. Classification of behavior in housed dairy cows using an accelerometer-based activity monitoring system. *Anim. Biotelemetry* **2015**, *3*, 15. [CrossRef]
6. Rahman, A.; Smith, D.; Little, B.; Ingham, A.; Greenwood, P.; Bishop-Hurley, G. Cattle behavior classification from collar, halter, and ear tag sensors. *Inf. Process. Agric.* **2018**, *5*, 124–133. [CrossRef]
7. Benaissa, S.; Tuyttens, F.; Plets, D.; Pessemier, T.; Trogh, J.; Tanghe, E.; Martens, L.; Vandaele, L.; Van Nuffel, A.; Joseph, W.; et al. On the use of on-cow accelerometers for the classification of behaviors in dairy barns. *Res. Vet. Sci.* **2019**, *125*, 425–433. [CrossRef] [PubMed]
8. Vázquez-Diosdado, J.A.; Paul, V.; Ellis, K.A.; Coates, D.; Loomba, R. A combined offline and online algorithm for real-time and long-term classification of sheep behaviour: Novel approach for precision livestock farming. *Sensors* **2019**, *19*, 3201. [CrossRef]
9. Bell, M.J.; Tzimiropoulos, G. Novel monitoring systems to obtain dairy cattle phenotypes associated with sustainable production. *Front. Sustain. Food Syst.* **2018**, *2*, 31. [CrossRef]
10. Zuko, M.; Jaja, I. Primiparous and multiparous Friesland, Jersey, and crossbred cows' behavior around parturition time at the pasture-based system in South Africa. *J. Adv. Vet. Anim. Res.* **2020**, *7*, 290–298. [CrossRef] [PubMed]
11. Rahman, A.N. Hormonal changes in the uterus during pregnancy—Lessons from the ewe: A review. *J. Agric. Rural Dev.* **2008**, *4*, 1–7. [CrossRef]
12. Huzzey, J.M.; von Keyserlingk, M.A.G.; Weary, D.M. Changes in feeding, drinking, and standing behavior of dairy cows during the transition period. *J. Dairy Sci.* **2005**, *88*, 2454–2461. [CrossRef]
13. Barrier, A.C.; Haskell, M.J.; Macrae, A.I.; Dwyer, C.M. Parturition progress and behaviours in dairy cows with calving difficulty. *Appl. Anim. Behav. Sci.* **2012**, *139*, 209–217. [CrossRef]
14. Jensen, M.B. Behaviour around the time of calving in dairy cows. *Appl. Anim. Behav. Sci.* **2012**, *139*, 195–202. [CrossRef]
15. Miedema, H.M.; Cockram, M.S.; Dwyer, C.M.; MacRae, A.I. Changes in the behaviour of dairy cows during the 24 h before normal calving compared with behaviour during late pregnancy. *Appl. Anim. Behav. Sci.* **2011**, *131*, 8–14. [CrossRef]
16. Arnold, G.W.; Morgan, P.D. Behaviour of the ewe and lamb at lambing and its relationship to lamb mortality. *Appl. Anim. Ethol.* **1975**, *2*, 25–46. [CrossRef]
17. Fogarty, E.S.; Swain, D.L.; Cronin, G.M.; Moreas, L.E.; Trotter, M. Can accelerometer ear tags identify behavioural changes in sheep associated with parturition? *Anim. Reprod. Sci.* **2020**, *216*, 106345. [CrossRef] [PubMed]
18. AFRC. *Energy and Protein Requirements of Ruminants: An Advisorymanual Prepared by the AFRC Technical Committee on Responses to Nutrients*; Cab International: Wallingford, VT, USA, 1993.

19. Smid, A.-M.C.; Weary, D.M.; von Keyserlingk, M.A.G. The influence of different types of outdoor access on dairy cattle behavior. *Front. Vet. Sci.* **2020**, *7*, 257. [CrossRef]
20. Black, R.A.; Krawczel, P.D. Effect of prepartum exercise on lying behavior, labor length, and cortisol concentrations. *J. Dairy Sci.* **2019**, *102*, 11250–11259. [CrossRef]
21. Bailey, D.W.; Trotter, M.G.; Knight, C.W.; Thomas, M.G. Use of GPS tracking collars and accelerometers for rangeland livestock production research. *Transl. Anim. Sci.* **2018**, *2*, 81–88. [CrossRef]
22. Munoz, C.; Campbell, A.; Hemsworth, P.; Doyle, R. Animal-Based Measures to Assess the Welfare of Extensively Managed Ewes. *Animals* **2018**, *8*, 2. [CrossRef]
23. Neethirajan, S. Transforming the adaptation physiology of farm animals through sensors. *Animals* **2020**, *10*, 1512. [CrossRef] [PubMed]
24. Alvarenga, F.A.P.; Borges, I.; Palkovič, L.; Rodina, J.; Oddy, V.H.; Dobos, R.C. Using a three-axis accelerometer to identify and classify sheep behaviour at pasture. *Appl. Anim. Behav. Sci.* **2016**, *183*, 104. [CrossRef]
25. Fogarty, E.S.; Swain, D.L.; Cronin, G.M.; Moreas, L.E.; Trotter, M. Behaviour classification of extensively grazed sheep using machine learning. *Comput. Electron. Agric.* **2020**, *169*, 105175. [CrossRef]

Article

Changes in Dairy Cow Behavior with and without Assistance at Calving

Bethan Cavendish [1], John McDonagh [2], Georgios Tzimiropoulos [3], Kimberley R. Slinger [1], Zoë J. Huggett [1] and Matt J. Bell [4,*]

1   School of Biosciences, Sutton Bonington Campus, University of Nottingham, Sutton Bonington LE12 5RD, UK; stybcc@nottingham.ac.uk (B.C.); kimberley.slinger@nottingham.ac.uk (K.R.S.); Zoe.Huggett3@nottingham.ac.uk (Z.J.H.)
2   School of Computer Science, Jubilee Campus, University of Nottingham, Nottingham NG8 1BB, UK; john.mcdonagh@nottingham.ac.uk
3   School of Electrical Engineering and Computer Science, Queen Mary University of London, London E1 4NS, UK; g.tzimiropoulos@qmul.ac.uk
4   Agriculture Department, Hartpury University, Gloucester GL19 3BE, UK
*   Correspondence: matt.bell@hartpury.ac.uk

Abstract: The aim of this study was to characterize calving behavior of dairy cows and to compare the duration and frequency of behaviors for assisted and unassisted dairy cows at calving. Behavioral data from nine hours prior to calving were collected for 35 Holstein-Friesian dairy cows. Cows were continuously monitored under 24 h video surveillance. The behaviors of standing, lying, walking, shuffle, eating, drinking and contractions were recorded for each cow until birth. A generalized linear mixed model was used to assess differences in the duration and frequency of behaviors prior to calving for assisted and unassisted cows. The nine hours prior to calving was assessed in three-hour time periods. The study found that the cows spent a large proportion of their time either lying (0.49) or standing (0.35), with a higher frequency of standing (0.36) and shuffle (0.26) bouts than other behaviors during the study. There were no differences in behavior between assisted and unassisted cows. During the three-hours prior to calving, the duration and bouts of lying, including contractions, were higher than during other time periods. While changes in behavior failed to identify an association with calving assistance, the monitoring of behavioral patterns could be used as an alert to the progress of parturition.

Keywords: dairy cows; behavior; birth; observations; management

Citation: Cavendish, B.; McDonagh, J.; Tzimiropoulos, G.; Slinger, K.R.; Huggett, Z.J.; Bell, M.J. Changes in Dairy Cow Behavior with and without Assistance at Calving. *Agriculture* 2021, 11, 722. https://doi.org/10.3390/agriculture11080722

Academic Editor: Manuel García-Herreros

Received: 15 June 2021
Accepted: 28 July 2021
Published: 29 July 2021

1. Introduction

There has been increased interest in the care and housing of cows with concerns for cow welfare given the increasing size of the average dairy herd across developed countries [1]. Animal welfare concerns are commonly directed at farm animals, and in particular housed and more intensive production systems with large numbers of animals [2]. With larger herds the expectation is often that each dairy stockperson will look after more animals as farms either seek to reduce labor costs or find it difficult to source skilled labor.

Close monitoring at calving is required by the stockperson to ensure the survival of the mother and her offspring, with problems potentially impacting on future lifetime performance. While some idea of expected calving date is often known, or estimated from time of insemination and gestation length, this estimate is often imprecise and requires some subjective judgement by the farmer with regular checks during late pregnancy to ensure a successful outcome for the mother and offspring. To assist a stockperson at calving, and given the importance of a successful birth and potential need for intervention, a number of sensor technologies have been developed. These technologies have largely been based on accelerometers and movement detection [3,4], or an alternative is computer

19

vision [5,6], which have been developed to support farm management and improve animal health and wellbeing, and ultimately productivity. The frequency of lying, standing and tail movements of an animal have been found to change in the period prior to calving in both dairy [7] and beef cattle [8], and may give some indication of the need for assistance. Dystocia is fairly common in dairy cows and is a major cause of calf mortality [9,10]. Barrier et al. [10] found that calves which survived dystocia had poorer welfare in the neonatal period and possibly beyond, with lower passive immunity transfer, higher mortality and higher indicators of physiological stress. Although preventing dystocia is close to impossible, quick and timely intervention will help avoid the risk of poor health outcomes. Individual evaluation and continuous monitoring of dairy cows around the time of calving is important to identify any need for intervention or health problems as early as possible. The impact on lifetime performance and labor cost is estimated to range from £110 to £400 per assisted calving [11].

The objective of this study was to characterize calving behavior of dairy cows and compare the duration and frequency of behaviors for assisted and unassisted dairy cows at calving. The hypothesize of the current study was that there would be a difference between the behavior of cows that were assisted and unassisted at calving, which could provide some insight for enhanced monitoring.

2. Materials and Methods

Approval for this study was obtained from the University of Nottingham animal ethics committee before commencement of the study (approval number 198).

2.1. Data

Video cameras (5 Mp, 30 m IR. Hikvision HD Bullet; Hangzhou, China) were used to record Holstein-Friesian dairy cows at the Nottingham University Dairy Centre (Sutton Bonington, Leicestershire, UK) prior to calving. Cameras were recording at 20 frames per second. Three calving pens with two surveillance cameras looking into each pen were used to obtain 24 h video footage of 35 individual cows between April and June 2018. Both cameras on each pen allowed full coverage of the area and were approximately at a 45-degree angle looking into the pen. Each calving pen housed a maximum of eight cows. At three weeks before expected calving, each cow was moved into one of the three calving pens so that the entire calving process could be closely monitored. Of the 35 cows monitored, 17 were primiparous and 18 multiparous. The need for a birth to be assisted was recorded for each cow and determined by the same experienced farm staff from visual assessment on calving progress. Cows were managed and housed within their normal environment.

2.2. Observations

The video recording for each cow was annotated from 9 h prior to giving birth by three observers using custom made scripts in PyTorch 1.5 framework to record the behavior profile of each cow with time. The start of the continuous observation period was determined as 9 h from when the calf was fully expelled at birth using the video recording, and considered a time when no visual signs of calving behavior are observed. A total of 19,191 individual behavioral observations were recorded from all 35 cows. To ensure accuracy of video behavior annotations, the video was segmented into short clips for each behavior, and all video clips subsequently checked for accuracy by one of the three observers. Seven behaviors were recorded, which were:

1. Standing: The cow is still on all four legs.
2. Lying: The midway transition of when the cow is about to lie down to when they start to rise again.
3. Walking: Movement of more than two steps.
4. Shuffle: Cow circles on the spot or moves slightly with a step or two.
5. Contractions: Visible straining while lying down.

6. Eating: Cow puts its head through the feeding barrier until the moment it pulls its head back out from the feeding barrier.
7. Drinking: Head is over the water trough and regular head movement towards the trough.

2.3. Statistical Analysis

For the analysis, the 9 h prior to giving birth was split into three-hour time periods, with period three ending with the birth. The duration of behaviors in seconds and frequency were determined for each time period. A total of 735 behavior records were obtained from 35 cows (35 × 7 behaviors × 3 time periods).

Behavior records were analyzed using a generalized linear mixed model in Genstat Version 19.1 (Lawes Agricultural Trust, 2018). A binomial error distribution and a logit link function added was fitted to the fixed effects of assistance, time period, behavior and parity for the dependent variables of duration and frequency of behaviors in Equation (1):

$$Y_{ijkl} = \mu + A_i \times T_j \times B_k + P_l + E_{ijkl} \tag{1}$$

where Y_{ijkl} is the dependent variable of behavior duration or frequency; μ = overall mean; A_i = fixed effect for assistance at calving (i = 0 for unassisted or 1 for assisted); T_j = fixed effect of time period (j = 1 to 3); B_k = fixed effect of behavior (k = standing, lying, walking, shuffle, contractions, eating and drinking); P_l = fixed effect of parity (l = primiparous or multiparous); E_{ijkl} = random error term.

The back-transformed predicted means for behavior duration and frequency were expressed as the proportion of total time or count during each time period. The three-hour time periods allowed analysis of the proportion of time and frequency of behaviors to be compared for several behaviors within a time period. Significance was attributed at $p < 0.05$.

3. Results

Of the 35 calvings, there were four primiparous and four multiparous calvings that required assistance by the farm stockperson, with all other calvings being unassisted. The study found no difference in duration or frequency of behaviors between cows that had an assisted or unassisted calving. Furthermore, there was no difference between primiparous and multiparous cows (Table 1). Differences were found in the duration of behaviors ($p < 0.001$) with the majority of time spent lying (0.49) or standing (0.35) with other behaviors being 0.04 or less across the 9 h studied (Figure 1). In the final three hours prior to calving, the proportion of time for lying and contractions increased and the time spent standing, drinking and eating decreased ($p < 0.001$; Table 1 and Figure 2).

Table 1. Effects of parity, time period and calving assistance on the duration of dairy cow ($n = 35$) behaviors as a proportion of time.

Variable		Mean (s.e.)	df	F Statistic	p Value
Parity	Primiparous	0.04 (0.01)	1	0.02	0.877
	Multiparous	0.04 (0.01)			
Time period [1]	Period 1	0.03 (0.02)	2	0.2	0.795
	Period 2	0.05 (0.01)			
	Period 3	0.05 (0.01)			
Assistance [2]	Assisted	0.05 (0.01)	1	0.2	0.631
	Unassisted	0.04 (0.01)			
Behavior [3]			6	130	<0.001
Time period × assistance	Period 1/Assisted	0.04 (0.02)	2	0.5	0.631
	Period 1/Unassisted	0.03 (0.03)			
	Period 2/Assisted	0.06 (0.02)			
	Period 2/Unassisted	0.04 (0.01)			
	Period 3/Assisted	0.04 (0.02)			
	Period 3/Unassisted	0.05 (0.01)			

Table 1. *Cont.*

Variable		Mean (s.e.)	df	F Statistic	*p* Value
Assistance × behavior	Assisted/Contractions	0.01 (0.01)	6	0.9	0.515
	Assisted/Drinking	0.01 (0.01)			
	Assisted/Eating	0.03 (0.02)			
	Assisted/Lying	0.47 (0.03)			
	Assisted/Shuffle	0.04 (0.01)			
	Assisted/Stand	0.36 (0.03)			
	Assisted/Walking	0.02 (0.01)			
	Unassisted/Contractions	0.002 (0.004)			
	Unassisted/Drinking	0.01 (0.01)			
	Unassisted/Eating	0.06 (0.01)			
	Unassisted/Lying	0.52 (0.01)			
	Unassisted/Shuffle	0.03 (0.01)			
	Unassisted/Stand	0.34 (0.01)			
	Unassisted/Walking	0.01 (0.003)			
Time period × behavior	Period 1/Contractions	0 (0.001)	12	4.4	<0.001
	Period 1/Drinking	0.01 (0.01)			
	Period 1/Eating	0.10 (0.02)			
	Period 1/Lying	0.47 (0.03)			
	Period 1/Shuffle	0.04 (0.01)			
	Period 1/Stand	0.37 (0.03)			
	Period 1/Walking	0.02 (0.01)			
	Period 2/Contractions	0.002 (0.002)			
	Period 2/Drinking	0.02 (0.01)			
	Period 2/Eating	0.08 (0.01)			
	Period 2/Lying	0.44 (0.03)			
	Period 2/Shuffle	0.04 (0.01)			
	Period 2/Stand	0.41 (0.03)			
	Period 2/Walking	0.02 (0.01)			
	Period 3/Contractions	0.07 (0.01)			
	Period 3/Drinking	0.004 (0.003)			
	Period 3/Eating	0.01 (0.01)			
	Period 3/Lying	0.58 (0.03)			
	Period 3/Shuffle	0.04 (0.01)			
	Period 3/Stand	0.28 (0.02)			
	Period 3/Walking	0.01 (0.01)			
Time period × behavior × assistance [4]			12	0.4	0.966

[1] Periods 1, 2 and 3 were observations 7 to 9, 4 to 6 and 1 to 3 h before calving, respectively. [2] Births were either assisted or unassisted by farm staff. [3] Predicted mean values shown in Figure 1. [4] Predicted mean values shown in Figure 2.

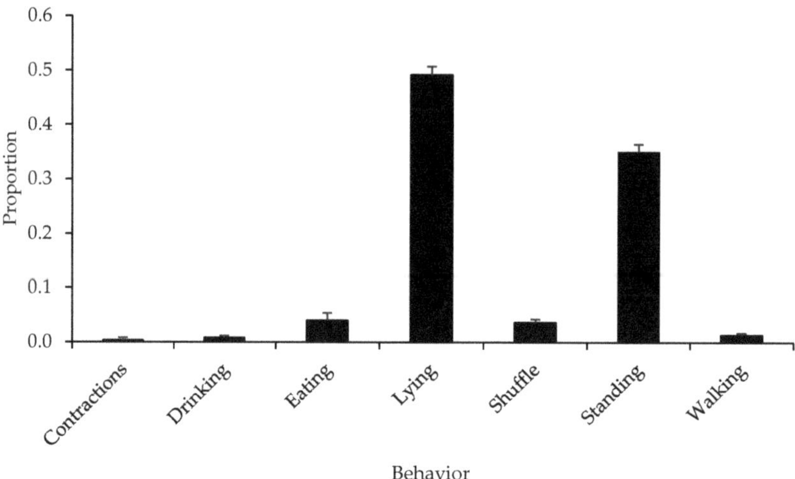

Figure 1. Predicted mean (± SEM) proportion of time dairy cows (*n* = 35) spent doing different behaviors during the 9 h prior to calving.

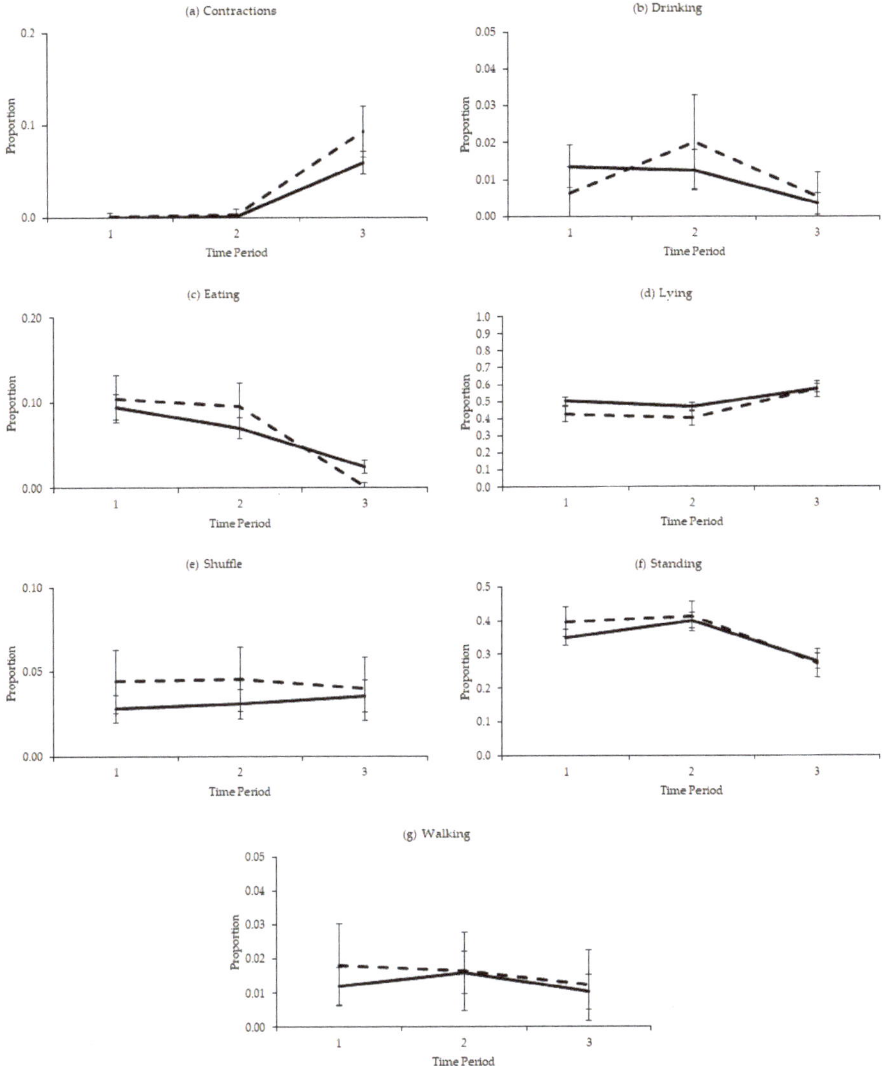

Figure 2. Predicted mean (± SEM) proportion of time that were (**a**) contractions, (**b**) drinking, (**c**) eating, (**d**) lying, (**e**) shuffle, (**f**) standing and (**g**) walking behavior for assisted (dashed line) or unassisted (solid line) dairy cow calvings (*n* = 35) in time periods one to three, with period three ending with the birth.

Differences were also found in the frequency of behaviors ($p < 0.001$) with standing (0.36) and shuffle (0.26) being most frequent, with other behaviors being 0.09 or less across the 9 h studied (Figure 3).

In the final three hours prior to calving, the frequency of lying and contraction bouts increased and the standing, shuffle, walking, drinking and eating bouts decreased ($p < 0.001$; Table 2 and Figure 4).

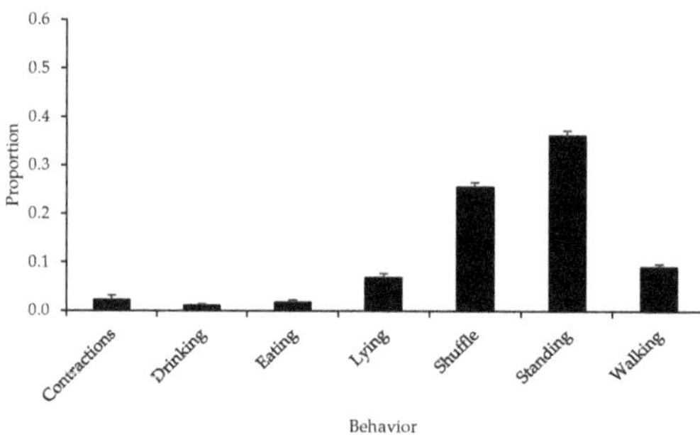

Figure 3. Predicted mean (±SEM) proportion of observations for different dairy cow (n = 35) behaviors during the 9 h prior to calving.

Table 2. Effects of parity, time period and calving assistance on the frequency of dairy cow (n = 35) behaviors as a proportion of observations.

Variable		Mean (s.e.)	df	F Statistic	p Value
Parity	Primiparous	0.06 (0.01)	1	0.00003	0.996
	Multiparous	0.06 (0.01)			
Time period [1]	Period 1	0.06 (0.01)	2	1.7	0.191
	Period 2	0.07 (0.01)			
	Period 3	0.06 (0.01)			
Assistance [2]	Assisted	0.07 (0.01)	1	0.2	0.630
	Unassisted	0.06 (0.01)			
Behavior [3]			6	140	<0.001
Time period × assistance	Period 1/Assisted	0.07 (0.01)	2	2.1	0.119
	Period 1/Unassisted	0.05 (0.02)			
	Period 2/Assisted	0.08 (0.01)			
	Period 2/Unassisted	0.07 (0.01)			
	Period 3/Assisted	0.05 (0.01)			
	Period 3/Unassisted	0.07 (0.01)			
Assistance × behavior	Assisted/Contractions	0.05 (0.01)	6	1.3	0.257
	Assisted/Drinking	0.01 (0.004)			
	Assisted/Eating	0.01 (0.01)			
	Assisted/Lying	0.07 (0.01)			
	Assisted/Shuffle	0.24 (0.02)			
	Assisted/Stand	0.34 (0.02)			
	Assisted/Walking	0.09 (0.01)			
	Unassisted/Contractions	0.01 (0.01)			
	Unassisted/Drinking	0.01 (0.002)			
	Unassisted/Eating	0.03 (0.004)			
	Unassisted/Lying	0.06 (0.01)			
	Unassisted/Shuffle	0.27 (0.01)			
	Unassisted/Stand	0.38 (0.01)			
	Unassisted/Walking	0.09 (0.01)			
Time period × behavior	Period 1/Contractions	0.002 (0.003)	12	40.2	<0.001
	Period 1/Drinking	0.02 (0.01)			
	Period 1/Eating	0.05 (0.01)			
	Period 1/Lying	0.03 (0.01)			
	Period 1/Shuffle	0.31 (0.02)			
	Period 1/Stand	0.45 (0.02)			
	Period 1/Walking	0.13 (0.02)			
	Period 2/Contractions	0.01 (0.01)			
	Period 2/Drinking	0.02 (0.01)			
	Period 2/Eating	0.04 (0.01)			
	Period 2/Lying	0.04 (0.01)			
	Period 2/Shuffle	0.31 (0.02)			
	Period 2/Stand	0.45 (0.02)			
	Period 2/Walking	0.13 (0.01)			
	Period 3/Contractions	0.28 (0.01)			
	Period 3/Drinking	0.004 (0.002)			
	Period 3/Eating	0.003 (0.002)			
	Period 3/Lying	0.29 (0.01)			
	Period 3/Shuffle	0.16 (0.01)			
	Period 3/Stand	0.21 (0.01)			
	Period 3/Walking	0.04 (0.01)			
Time period × behavior × assistance [4]			12	0.5	0.926

[1] Periods 1, 2 and 3 were observations 7 to 9, 4 to 6 and 1 to 3 h before calving, respectively. [2] Births were either assisted or unassisted by farm staff. [3] Predicted mean values shown in Figure 3. [4] Predicted mean values shown in Figure 4.

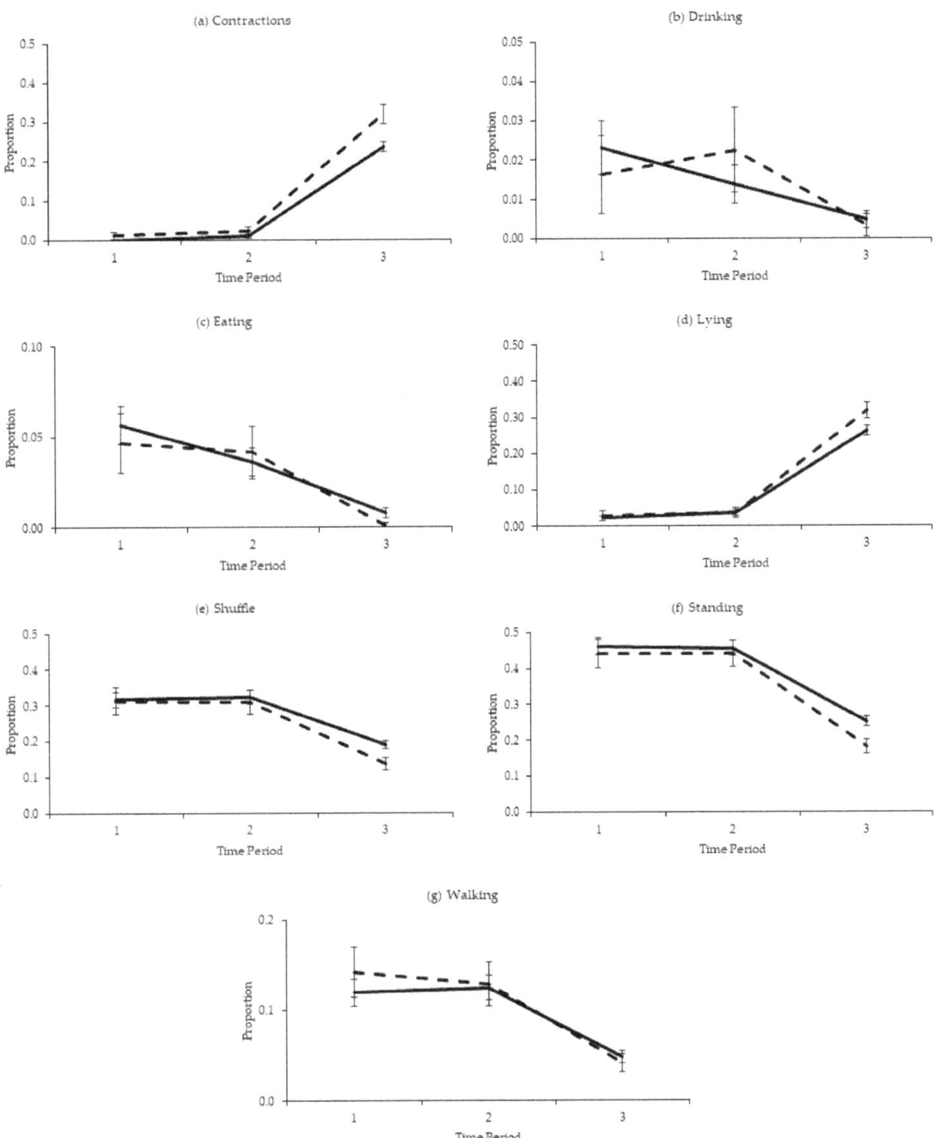

Figure 4. Predicted mean (± SEM) proportion of observations that were (**a**) contractions, (**b**) drinking, (**c**) eating, (**d**) lying, (**e**) shuffle, (**f**) standing and (**g**) walking behavior for assisted (dashed line) or unassisted (solid line) dairy cow calvings ($n = 35$) in time periods one to three, with period three ending with the birth.

4. Discussion

This study found that when monitoring calving the duration and frequency of lying and contraction bouts increased in the last three hours prior to birth compared to other time periods studied. Observing contractions and their increased frequency, along with increased frequency and time spent lying, can be used as indicators of progress in parturition. During the nine hours studied prior to calving, cows spent a large proportion of their time either lying or standing in their late pregnancy. No difference in behavioral patterns were found between assisted and unassisted calvings in the current study, but have been found by others [7,12]. The failure to find a difference between assisted and unassisted calvings may have been influenced by observations being conducted until the calf was fully expelled and any assistance being subjectively determined by farm staff. Additionally, only 23% (8 of the 35 cows) in the current study needed assistance when calving, and therefore further observations of assisted births would add to the study.

Cows tend to be lying when contractions are occurring. The behavior of dairy cows has been heavily researched due to concerns to dairy cow welfare and as an indicator of poor health [13]. Lying is a highly motivated behavior in dairy cows, with cows prioritizing lying over other behaviors such as feeding, and especially after a period when these behaviors have been limited [13]. Typically, cows when indoors will spend between 10–12 h per day lying, and between eight and 10 h per day when grazing [14]. The difference may reflect more time needed for eating and walking when at pasture. In the current study cows spent about 12 h per day lying and eight hours standing, with more time spent lying and less time standing, drinking and eating as parturition progressed. Miedema et al. [7] found the frequencies of lying and tail raising increased in the final six-hours before calving and that changes in standing and lying could potentially be used as a predictor of calving. The findings of the current study would also support the use of lying and standing transitions as a means for farmers and technology to detect the progress of parturition and imminent birth. Giaretta et al. [4] also found increased tail movements as an important indicator of calving progress, along with decreased eating behavior and rumination time. Further behaviors such as tail movements and rumination time may have added to the current study since they are potentially visible on video footage.

Schuenemann et al. [12] suggested that dystocic births are characterized by an increase in abdominal contractions for around 95 min until intervention is required. Therefore, if contractions can be tracked accurately, and potentially with technology, a prediction of dystocia could potentially be made given its importance in the monitoring of parturition. Electronic devices such as abdominal belts or intravaginal thermometers to detect uterine contractions and body temperature changes have been proposed as potential solutions [3,4]. Furthermore, the live video feed could be monitored using camera surveillance software to track individual cow behavior since the cows are often indoors when calving [5,6]. Several studies have proposed sensor technology for classifying cow behavior [15–17]. During the study, changes in behavior were largely associated with standing, shuffle, walking and lying, with bouts of lying increasing in the period prior to calving, which is consistent with other studies [18]. This potential restlessness is known to relate to discomfort in animals and may reflect late stages of pregnancy or boredom [19].

5. Conclusions

No differences among assisted and unassisted calvings were found. During the 9 h that were studied before calving, cows spent a large proportion of their time either lying or standing, with a higher frequency of standing and shuffle bouts than other behaviors. The increased time and bouts of lying, including contractions, during the last three-hours prior to calving, and the reduction in other behaviors, could provide a means of tracking the progress of parturition by the stockperson and using technology. These findings could help support dairy farmers at calving and increase the welfare of cows and their offspring, and subsequent lifetime performance.

Author Contributions: Conceptualization, M.J.B. and G.T.; methodology, M.J.B. and G.T.; software, J.M.; validation, J.M.; formal analysis, B.C. and M.J.B.; investigation, B.C. and M.J.B.; resources, M.J.B.; data curation, M.J.B., K.R.S., Z.J.H. and J.M.; writing—original draft preparation, B.C. and M.J.B.; writing—review and editing, B.C. and M.J.B.; visualization, J.M.; supervision, M.J.B.; project administration, M.J.B.; funding acquisition, M.J.B. and G.T. All authors have read and agreed to the published version of the manuscript.

Funding: This research was funded by the Douglas Bomford Trust, the Engineering and Physical Sciences Research Council and the Biotechnology and Biological Sciences Research Council.

Institutional Review Board Statement: The study was conducted according to the guidelines of the Declaration of Helsinki, and approved by the Animal Ethics Committee at the University of Nottingham (approval number 198, 2018).

Informed Consent Statement: Not applicable.

Data Availability Statement: The analyzed datasets are available from the corresponding author on request.

Conflicts of Interest: The authors declare no conflict of interest.

References

1. Barkema, H.W.; Von Keyserlingk, M.A.G.; Kastelic, J.P.; Lam, T.J.G.M.; Luby, C.; Roy, J.P.; Leblanc, S.J.; Keefe, G.P.; Kelton, D.F. Invited review: Changes in the dairy industry affecting dairy cattle health and welfare. *J. Dairy Sci.* **2015**, *98*, 7426–7445. [CrossRef] [PubMed]
2. Lawrence, A.; Stott, A. Profiting from animal welfare: An animal-based perspective. *J. R. Agric. Soc. Engl.* **2009**, *170*, 40–47.
3. Rutten, C.J.; Kamphuis, C.; Hogeveen, H.; Huijps, K.; Nielen, M.; Steeneveld, W. Sensor data on cow activity, rumination, and ear temperature improve prediction of the start of calving in dairy cows. *Comput. Electron. Agric.* **2017**, *132*, 108–118. [CrossRef]
4. Giaretta, E.; Marliani, G.; Postiglione, G.; Magazzù, G.; Pantò, F.; Mari, G.; Formigoni, A.; Accorsi, P.A.; Mordenti, A. Calving time identified by the automatic detection of tail movements and rumination time, and observation of cow behavioral changes. *Animal* **2021**, *15*, 100071. [CrossRef] [PubMed]
5. Cangar, Ö.; Leroy, T.; Guarino, M.; Vranken, E.; Fallon, R.; Lenehan, J.; Mee, J.; Berckmans, D. Automatic real-time monitoring of locomotion and posture behavior of pregnant cows to calving using online image analysis. *Comput. Electron. Agric.* **2008**, *64*, 53–60. [CrossRef]
6. Bell, M.J.; Tzimiropoulos, G. Novel monitoring systems to obtain dairy cattle phenotypes associated with sustainable production. *Front. Sustain. Food Syst.* **2018**, *2*, 31. [CrossRef]
7. Miedema, H.M.; Cockram, M.S.; Dwyer, C.M.; MacRae, A.I. Changes in the behaviour of dairy cows during the 24h before normal calving compared with behaviour during late pregnancy. *Appl. Animal Behav. Sci.* **2011**, *131*, 8–14. [CrossRef]
8. Hyslop, J.; Ross, D.; Bell, M.; Dwyer, C. Observations on the time course of calving events in unassisted multiparous spring calving suckler cows housed in a straw bedded yard. In *Proceedings of the British Society of Animal Science*; Cambridge University Press: Cambridge, UK, 2008; p. 158.
9. Lombard, J.E.; Garry, F.B.; Tomlinson, S.M.; Garber, L.P. Impacts of Dystocia on Health and Survival of Dairy Calves. *J. Dairy Sci.* **2007**, *90*, 1751–1760. [CrossRef] [PubMed]
10. Barrier, A.C.; Haskell, M.J.; Birch, S.; Bagnall, A.; Bell, D.J.; Dickinson, J.; Macrae, A.I.; Dwyer, C.M. The impact of dystocia on dairy calf health, welfare, performance and survival. *Vet. J.* **2013**, *195*, 86–90. [CrossRef] [PubMed]
11. McGuirk, B.J.; Forsyth, R.; Dobson, H. Economic cost of difficult calvings in the United Kingdom dairy herd. *Vet. Rec.* **2007**, *161*, 685–687. [CrossRef] [PubMed]
12. Schuenemann, G.M.; Nieto, I.; Bas, S.; Galvão, K.N.; Workman, J. Assessment of calving progress and reference times for obstetric intervention during dystocia in Holstein dairy cows. *J. Dairy Sci.* **2011**, *94*, 5494–5501. [CrossRef] [PubMed]
13. Weary, D.M.; Huzzey, J.M.; von Keyserlingk, M.A.G. Board-invited review: Using behavior to predict and identify ill health in animals. *J. Anim. Sci.* **2009**, *87*, 770–777. [CrossRef] [PubMed]
14. Smid, A.-M.C.; Weary, D.M.; von Keyserlingk, M.A.G. The influence of different types of outdoor access on dairy cattle behavior. *Front. Vet. Sci.* **2020**, *7*, 257. [CrossRef] [PubMed]
15. Diosdado, J.; Barker, Z.; Hodges, H.; Amory, J.; Croft, D.; Bell, N.; Codling, E. Classification of behavior in housed dairy cows using an accelerometer-based activity monitoring system. *Anim. Biotelemetry* **2015**, *3*, 15. [CrossRef]
16. Rahman, A.; Smith, D.; Little, B.; Ingham, A.; Greenwood, P.; Bishop-Hurley, G. Cattle behavior classification from collar, halter, and ear tag sensors. *Inf. Process. Agric.* **2018**, *5*, 124–133. [CrossRef]
17. Benaissa, S.; Tuyttens, F.; Plets, D.; Pessemier, T.; Trogh, J.; Tanghe, E.; Martens, L.; Vandaele, L.; Van Nuffel, A.; Joseph, W.; et al. On the use of on-cow accelerometers for the classification of behaviors in dairy barns. *Res. Vet. Sci.* **2019**, *125*, 425–433. [CrossRef] [PubMed]
18. Huzzey, J.M.; von Keyserlingk, M.A.G.; Weary, D.M. Changes in Feeding, Drinking, and Standing Behavior of Dairy Cows during the Transition Period. *J. Dairy Sci.* **2005**, *88*, 2454–2461. [CrossRef]
19. Munksgaard, L.; Jensen, M.B.; Pedersen, L.J.; Hansen, S.W.; Matthews, L. Quantifying behavioural priorities—Effects of time constraints on behaviour of dairy cows, Bos taurus. *Appl. Anim. Behav. Sci.* **2005**, *92*, 3–14. [CrossRef]

Review

Proximal Sensing in Grasslands and Pastures

Shayan Ghajar * and Benjamin Tracy

School of Plant and Environmental Sciences, Virginia Tech, Blacksburg, VA 24061, USA; bftracy@vt.edu
* Correspondence: sghajar@vt.edu

Abstract: Reliable measures of biomass, species composition, nitrogen status, and nutritive value provide important indicators of the status of pastures and rangelands, allowing managers to make informed decisions. Traditional methods of sample collection necessitate significant investments in time and labor. Proximal sensing technologies have the potential to collect more data with a smaller investment in time and labor. However, methods and protocols for conducting pasture assessments with proximal sensors are still in development, equipment and software vary considerably, and the accuracy and utility of these assessments differ between methods and sites. This review summarizes the methods currently being developed to assess pastures and rangelands worldwide and discusses these emerging technologies in the context of diffusion of innovation theory.

Keywords: proximal; sensing; LiDAR; photogrammetry; grasslands; pastures

Citation: Ghajar, S.; Tracy, B. Proximal Sensing in Grasslands and Pastures. *Agriculture* **2021**, *11*, 740. https://doi.org/10.3390/agriculture11080740

Academic Editor: Matt J. Bell

Received: 7 July 2021
Accepted: 31 July 2021
Published: 4 August 2021

Publisher's Note: MDPI stays neutral with regard to jurisdictional claims in published maps and institutional affiliations.

1. Introduction

Grasslands cover an estimated 40% of the Earth's surface [1], performing vital ecosystem services such as cycling nutrients, carbon, and water, and providing habitat for wildlife and pasture for livestock [2]. However, grasslands are declining globally in extent, due to land-use changes such as development, conversion to cropland, or abandonment [1,3] or degrading due to climate change [4]. Assessing the health of a grassland necessitates the collection of reliable data on its productivity, plant species composition, and potentially its nutritive value [5]. Traditional techniques for collecting these data are time and labor-intensive, requiring field visits, specialized equipment, and teams with expert knowledge of local ecosystems. Advances in sensor technologies may reduce the time or labor required to conduct such measurements. While recent applications of remote sensing technologies in grasslands and pastures have been discussed in detail in prior research [6], proximal sensing technologies such as handheld sensors or sensors mounted on unmanned aerial vehicles (UAVs) flown at low altitude have received less attention.

Innovation diffusion is the study of how novel and potentially useful technologies spread throughout a social system [7]. Innovation diffusion theories emerged out of studies of the dissemination of novel agricultural practices and technologies [8]. As such, they may be superior frameworks by which to evaluate the spread of innovative proximal sensing technologies for measuring or predicting indicators of pasture or rangeland health and function.

To diffuse successfully, an innovation must have five attributes demonstrating its advantage over previous systems: relative advantage, definable as the perceived superiority of the innovation over other methods; compatibility with the technological, cultural, social, and economic systems into which it must be integrated; a complexity not exceeding that of necessity; trialability, which is a term for the ease by which the innovation may be tested without major commitment; and observability of the superior outcomes of the innovation [7].

In this paper, we assess the current state of proximal sensing technologies as they relate to measuring plant height and biomass, species composition, and nutritive value of pastures and rangelands and assess the advantages and obstacles to their adoption within the framework of theories on the diffusion of innovations.

2. Measuring Plant Height and Predicting Aboveground Biomass

Measurements of sward height and biomass can provide data on the structure and health of a grassland. Sward height measurements can indicate the vigor, habitat value, or maturity of a grassland. Similarly, aboveground biomass is an indicator of grassland or rangeland health and, when monitored long-term, can indicate whether a grassland is gaining, losing, or maintaining its capacity to provide valuable ecosystem services such as forage production or preservation of water quality [9]. Data on aboveground biomass allow managers to determine appropriate levels of forage utilization to meet economic goals while maintaining ecosystem services [10].

Traditional methods of measuring aboveground biomass rely on destructive harvest of forage from small areas delineated by a quadrat or frame [11]. The harvested biomass then must be dried in a dedicated oven before dry matter can be determined. This process is labor-intensive, and the ovens needed for drying samples can be a significant investment. On extensive rangelands, an additional limitation of this method is that samples are often collected along a small transect that must represent areas of hundreds or thousands of hectares in size due to limitations on time and labor available to technicians or scientists. As such, improper site selection can mean that the calculated aboveground biomass is not representative of the area, and management decisions based on the results may not have the desired effect [12].

2.1. Traditional Alternatives to Destructive Harvest

A number of alternatives to hand-harvesting forage samples have been developed to shorten the time required to assess biomass in a pasture. Plate meters are used for quick, non-destructive estimations of pasture biomass by measuring the height of a sward under compression by a plate on a measuring stick or pole. Plate meters are calibrated by taking measurements in a pasture typical of the area in which they are to be used, harvesting biomass on the site of each measurement, and comparing plate height to biomass with a regression model to develop a predictive equation [13]. Plate meters vary in accuracy depending on site and season, as morphology and stage of growth influence a tiller's resistance to compression [14]. Plate meters cannot be used if pasture species have robust reproductive tillers or stems that hold the plate above the sward canopy, as this will lead to inaccurate biomass predictions when plates are calibrated on vegetative swards [13]. Additionally, while the shortened sampling time compared to hand harvesting may allow for more sites to be sampled, the number of samples and area that can be sampled is still smaller than what can be accomplished with UAVs [15].

Robel poles are another non-destructive method of estimating pasture or rangeland biomass [16]. A Robel pole is a rod marked with contrasting colors at set intervals that is placed vertically in a sward. Personnel then stand at a set distance from the pole and record the height at which the sward obscures its markings. This is performed from four directions (often the cardinal directions) per placement, and the process is repeated until the necessary sample size has been collected. Robel poles avoid many of the issues intrinsic to plate meters by providing an indication of vegetative height and density without the need for compressing tillers. However, they face many of the same logistical limitations: a significant commitment of personnel and time to measure relatively small areas of often large ecosystems.

2.2. Use of Proximal Sensors to Measure Plant Height and Predict Biomass

Methods for assessing pasture height and biomass using proximal sensors vary. However, the principle for most methods is similar to that of plate meters in that pasture height measurements are initially compared to hand-harvested biomass to calibrate a predictive equation [15]. Photographs, LiDAR, and ultrasonic measurements are the most common methods used in proximal sensing pasture assessments [17,18].

Photographic determinations of pasture height may be carried out with a simple red-green-blue (RGB) camera mounted onto a UAV or other vehicle [19] or held on a pole [20]. Ground control points (GCPs) are high-visibility objects used to provide a frame of reference for determining the position of the UAV in the pasture being photographed and are georeferenced using GPS after they are placed in the pasture [21]. The UAV is flown over the pasture, taking pictures rapidly at low altitude. The pictures are then loaded into a Structure from Motion (SfM) photogrammetry program which uses the GCP to reconstruct the path of the UAV and generate a point cloud, from which a digital surface model (DSM) and digital elevation model (DEM) are derived (Figure 1). By subtracting the height of the DEM from the DSM, canopy height may be calculated for a given point [21]. As with plate meters, these height data are then combined with hand-harvested biomass to derive a regression model to predict pasture biomass as a function of UAV-measured height [22]. Once the model has been developed and tested, the UAV's height data may be used for determining biomass without the need for further hand-harvesting of biomass samples [21].

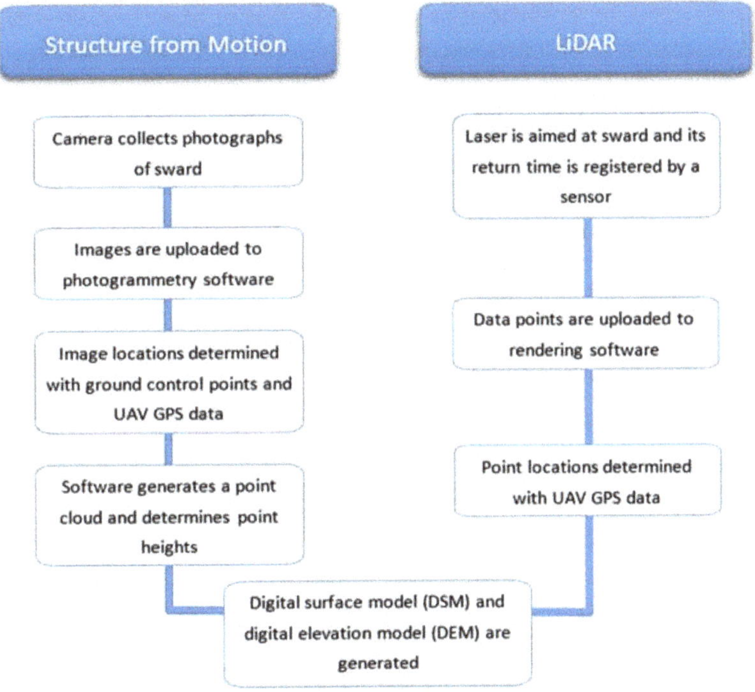

Figure 1. The processes by which pasture height data are generated using Structure for Motion photogrammetry and LiDAR in proximal sensing.

Light detection and ranging (LiDAR) generates point clouds much like SfM techniques but relies on a different mechanism to do so. With LiDAR, a laser aimed at the ground bounces back to a sensor, which records the time it took for the light to return, and from this interval calculates the distance from the LiDAR equipment [20].

When combined with the position of the UAV, LiDAR data can provide thousands of height measurements of a pasture or rangeland with a single flight [23]. Once the point cloud has been generated, a DSM is developed. The calibration of biomass prediction from height data is similar to the process used in SfM, with an initial comparison of LiDAR-generated height data with hand-harvested biomass samples [20].

Ultrasonic sensors operate similarly to LiDAR in that they emit a signal and calculate sward height based on the time it takes the signal to return to the sensor—the primary difference between the two being that one sensor uses ultrasound, and the other, light [24]. Ultrasonic sensors are mounted to vehicles both to facilitate more speedy measures and because the height of the sensor relative to the ground needs to remain constant for the sensor to correctly calculate sward height [17]. Recent improvements in ultrasonic sensor technology allow the sensor to detect the distance of both the ground and the sward canopy, resulting in more accurate measurements of height, as well as predictions of biomass [25]. However, the improved sensors are not yet commercially available.

Height measurements generated by proximal sensors can also be combined with spectral reflectance measurements to more accurately estimate pasture biomass [26]. Spectral reflectance indices are a measurement of the reflectance of objects at specific wavelengths. In a pasture setting, spectral reflectance is influenced by a number of factors, including chlorophyll concentration in leaves, plant species in the pasture, the presence and proportion of senescent leaves, and the leaf area index (LAI) of the pasture, which is the ratio of leaf area per given unit of ground area [27]. Spectral reflectance indices have been used on their own to estimate biomass in pastures and rangelands, but the many factors influencing reflectance at different wavelengths lead to difficulties in standardizing approaches between sites. Additionally, as the LAI increases, the sensitivity of reflectance indices decreases as less light is reflected back from denser swards [17]. In practical terms, this means that the most productive pastures present the greatest difficulty in accurately estimating biomass using spectral reflectance indices.

However, when spectral reflectance is combined with LiDAR or ultrasonic measurements of pasture height in a linear model, predictions of pasture biomass can be more accurate than by using either technique alone (Table 1). LiDAR combined with the NDVI resulted in a 25% increase in prediction accuracy for green pasture biomass [27]. This is a result of the capacity of spectral reflectance to differentiate between green biomass and senescent material, while LiDAR measurements can detect only standing biomass regardless of stage of growth [26]. Similarly, combining ultrasonic height measurements with spectral measurements such as the normalized spectral vegetation index (NSVI) can result in higher accuracy for predicting biomass than using either technology alone [26].

Table 1. Studies using proximal sensing technologies to estimate pasture height or dry matter (DM) yield.

Author(s)	Year	Technology Used [1]	Variable	R^2	Pasture Type	Location
Frick and Wachendorf [26]	2013	Combination of ultrasonic and spectral sensors	DM	0.83 in mixed swards 0.88–0.90 for species-specific calibrations	Mixed grass and legume swards	Germany
Schaefer and Lamb [27]	2016	Combination of LiDAR and spectral sensors	Green DM	0.61	Tall fescue (*Schedonorous arundinacea*)	Australia
Cooper et al. [28]	2017	SfM, TLS	DM	SfM: 0.72 TLS: 0.57	Smooth brome (*Bromus inermis*)	South Dakota, USA
Wang et al. [23]	2017	LiDAR	DM	0.34	Semi-arid steppe	Inner Mongolia, China
Legg and Bradley [25]	2019	Ultrasonic sensor	DM	0.7–0.8	Unspecified	New Zealand
Grüner et al. [21]	2019	SfM	DM Height	0.64–0.75 0.59–0.81	Mixed grass and legume swards	Germany
Obanawa et al. [20]	2020	SfM from pole-mounted and UAV-mounted cameras; LiDAR	Height	0.94 for UAV SfM, 0.91 for handheld SfM, 0.93 for LiDAR	Annual ryegrass (*Lolium multiflorum*)	Japan

[1] Abbreviations: LiDAR: light detection and ranging; SfM: structure from motion photogrammetry; TLS: terrestrial laser scanning; UAV: unmanned aerial vehicle.

Structure from Motion, LiDAR, and ultrasonic sensor technologies generate useful predictions of forage height and biomass using current techniques. However, each has unique advantages and disadvantages. The primary advantage of LiDAR is its accuracy. Obanawa and colleagues reported a margin of error for measured pasture height of 12 ± 10 mm for LiDAR compared to 24 ± 13 mm for SfM [20]. Additionally, ultrasonic sensors usually assume a fixed distance to the ground due to being mounted on a vehicle, yet terrain and bumpy driving may interfere with the validity of that assumption [17]. However, LiDAR equipment can be prohibitively expensive compared to the equipment needed for ultrasonic measurements or SfM. Another consideration is weather—SfM may be confounded by days with highly variable light conditions, whereas LiDAR and ultrasonic technologies are more robust to such situations. In summary, SfM and ultrasonic approaches may be more accessible for many people due to the simplicity and cost of equipment, but for height and yield estimations, LiDAR can be more accurate.

3. Grassland Species Composition

Determining the composition of plant communities in pastures and rangelands is critical to assessing the ecological health of a landscape and making informed management decisions. A census of plant species composition can identify species of economic value, conservation concern, or undesirable species which may need suppression. Regular monitoring of plant species composition can provide data on shifts in plant communities resulting from past management, biotic and abiotic stressors, or disturbances such as fire or drought [5]. Most surveys of plant species composition in pastures or rangeland require the same dedication of time and labor as the assessments of biomass discussed above and also face the same limitations in sampling area and representativeness of sites selected for survey. However, measuring plant species composition also requires personnel with proficiency in plant identification and familiarity with species local to sampling sites. The seasonality of plant growth may also limit preferable survey seasons to certain times of the year. Consequently, time and personnel availability remain substantial constraints to the number and scope of surveys that can be conducted on any particular landscape. Developing more efficient survey methods could increase the scope and frequency of monitoring programs.

Surveys of species composition using UAVs rely on the collection of high-resolution imagery for plant identification, followed by data processing and analysis. High-resolution imagery can be collected with a variety of cameras mounted to UAVs, including unmodified commercially available RGB cameras, or cameras specifically designed or modified to capture light outside the visible range, such as NIR [29]. After images have been collected, they may be evaluated to determine if blurring, overexposure, or other issues that preclude reliable analysis are present. Remaining images are then processed with software, which uses the timestamp and GPS data associated with each photo to create an orthomosaic—a single image stitched together from the many collected by the drone [30]. The orthomosaic may require additional processing prior to final analyses depending on the site, equipment, software, and species being surveyed. Methods of analysis can then include visual tagging of species by human operators or using mapping or image processing software to identify objects or areas based on patterns of pixels in the orthomosaic [31].

Past research using UAVs to identify plants at the species level have largely focused on a single highly visible species, such as tree species [32,33], weeds in monocropped farming systems [34], woody invasive species [35], species with conspicuous blooms [36], or species in arid landscapes with unique morphology [37]. These studies consistently found that analyses of UAV-derived high-resolution imagery compared favorably with traditional sampling methods when assessing the abundance and distribution of a target species.

We found only one study using UAV-derived high-resolution imagery to classify multiple species simultaneously in a grassland setting. Lu and He used blue, green, and NIR spectrum imagery processed with machine learning software to assess the abundance and distribution of six species in a temperate grassland, including forbs and grasses [29]. The simultaneous assessment necessitated training the image analysis program to recognize the species. This was achieved by identifying species in the field and photographing them with the equipment to be used in the UAV surveys in order to generate differentiable reflectance values for spectral analysis. This approach averaged 85% accuracy when compared with on-ground assessments; however, the authors noted that the method failed to detect plants below the sward canopy, whether from lodging or early growth stage. Additionally, due to differences in spectral reflectance species had at different stages of their life cycles, it was necessary for the authors to sample some species multiple times during the growing season to retrain the software.

Though comprehensive species inventories remain beyond the capabilities of proximal sensing technology at present, UAV-derived high-resolution imagery has been used to map cover classes in a variety of environments: desert ecosystems in West Asia [38], Western rangelands of North America [30,39–41], and grasslands in the Tibetan plateau [42]. The processes are similar to those outlined above for species identification but with lower-resolution imagery and training personnel or programs to recognize vegetation types rather than species based on pixel patterns [30].

Thus far, plant identification and mapping using proximal sensing technology is incapable of replacing field work for comprehensive inventories of species on most landscapes. Even when multiple species may be identified using sensor imagery, species lower in the sward canopy may go undetected, and a field crew proficient in plant identification is necessary for helping train image processing software to recognize species of interest. Nevertheless, proximal sensing technologies have demonstrated their utility in expanding the scope and frequency of monitoring programs in many contexts, as well as their usefulness in collecting data about large shifts across landscapes. As such, they can readily augment, rather than replace, traditional methods of conducting plant surveys for scientists and land managers.

4. Nutritional Status and Nutritive Value

The process for estimating the nutritional status of plants—e.g., what concentrations of nitrogen or phosphorus they may have—is functionally identical to the process for estimating nutritive value of the sward for grazing animals. For both, a proximal sensor collects images of the site across a range of spectra, while a field crew collects samples of pasture from geotagged sites in the field. The samples are chemically analyzed for the variables of interest, and then these values are compared to spectral data by means of linear regression and/or machine learning to find the spectrum and equation that best explain variance in the variable of interest [43].

Results using this technique vary, with high reported coefficients of determination for some monospecific stands of forage [40] and more variable predictive capability in heterogeneous swards [44] (Table 2). Approaches combining spectral data with other environmental variables reported higher prediction accuracy than those relying on spectral data alone, both in homogenous and heterogenous swards. A study of nutritive value of alfalfa (*Medicago sativa*) in the United States reported 9–17% higher prediction accuracy when combining spectral data and growing degree units than when models were based on spectral data alone [45]. Another study on mixed temperate grasslands in Germany reported a 21% increase in prediction accuracy for crude protein (CP) and 91% increase for prediction accuracy of acid detergent fiber (ADF) when ultrasonic sward height measurements were included in calibration models than with spectral data alone [46]. Similarly, a study on perennial ryegrass (*Lolium perenne*) and white clover (*Trifolium repens*) swards in New Zealand reported 21% greater prediction accuracy for CP and 28% greater prediction accuracy for metabolizable energy (ME) when spectral data were combined with selected topographic and edaphic variables in models [47].

Table 2. Studies using proximal sensing technologies to assess forage nutrient concentrations in pastures or rangelands.

Author(s)	Year	Technology Used	Variables [1]	R^2	Pasture Type	Location
Safari et al. [46]	2016	Handheld spectroradiometer and ultrasonic sensor	CP	0.64–0.85	Mixed temperate grasslands	Germany
			ADF	0.63–0.75		
Noland et al. [45]	2018	Near-infrared sensor	CP	0.87	Alfalfa (*Medicago sativa*) pasture	Minnesota, USA
			NDF	0.76		
Serrano et al. [48]	2018	Near-infrared sensor and optical sensor	CP	0.69	Dryland Meditteranean pastures	Portugal
			NDF	0.78		
Pullanagari et al. [47]	2018	Hyperspectral sensors and topographic data	CP	0.83	Perennial ryegrass (*Lolium perenne*) and white clover (*Trifolium repens*) sward	New Zealand
			NDF	0.76		
Wijesingha et al. [44]	2020	Hyperspectral sensors	CP	0.42	Mixed temperate grasslands	Germany
			ADF	0.34		
Michez et al. [43]	2020	Optical and multispectral sensors	CP	0.54	Timothy (*Phleum pratense*) pasture	Belgium
			ADF	0.82		
			NDF	0.84		
Duranovich et al. [49]	2020	Hyperspectral sensors	CP	0.78	Perennial ryegrass and white clover sward	New Zealand
			ADF	0.55		
			NDF	0.54		
			ME	0.67		

[1] Abbreviations: CP: crude protein. ADF: acid detergent fiber. NDF: neutral detergent fiber. ME: metabolizable energy.

5. Innovation Diffusion and Proximal Sensing Technologies

Returning to innovation diffusion theory, an innovation must have five characteristics if it is to be widely adopted throughout a system: relative advantage over past methods; compatibility with needs and values of potential adopters, as well as extant social, technological, and economic systems; complexity not exceeding that of necessity; trialability, or capacity to be tested before committing to the innovation; and observability of the superior outcomes of the innovation [7].

5.1. Relative Advantage

Proximal sensing technologies have the capacity to generate vastly more data than traditional methods of pasture and rangeland assessment. Additionally, proximal sensors mounted on UAVs can travel farther, faster than most fieldwork teams. For certain aspects of assessing pasture or rangeland health, such as measuring height and productivity, proximal sensing technology has a high degree of accuracy and clear relative advantage over repeatedly hand-harvesting pasture. In other areas, however, such as comprehensive assessments of pasture species, or measuring the nutritive value or nutrient status of a sward, proximal sensing technology can augment but not replace traditional methods of data collection at present.

5.2. Compatibility

Compatibility refers to the extent to which an innovation aligns with the needs, values, and past experiences of potential adopters, as well as the extent to which it can be integrated into extant social, economic, and technological systems [7]. Scientists and natural resources professionals both need and value enhanced capacity for data collection [50]. Similarly, farmers and ranchers face a trend of increasing intensification in agricultural production, with an increased focus on data collection and precision agriculture [51]; however, in some countries, the median age of agricultural producers is over 60 years [52], and age has been shown to have an inverse relationship with willingness to adopt technological innovations in some agricultural communities [53,54].

While proximal sensing technologies are in the process of emerging from and being further integrated into extant economic and technological systems, prior research on analogous precision agriculture innovations demonstrates that a threshold of adoption may need to be reached before industry-scale informational and technological support for potential adopters is available [55]. At present, there are few studies on the rates of adoption of proximal sensing technologies or related technologies such as UAVs by agricultural producers, and extant studies focus primarily on row crop production rather than pasture [56]. As such, it seems unlikely that the threshold of adoption of proximal sensing technologies is high enough for farmers and ranchers to expect consistent commercial support for proximal sensing technologies they may wish to integrate into their farms.

5.3. Complexity

The complexity of various proximal sensing technologies is likely to be a barrier to adoption for many land managers such as farmers or conservation professionals. Traditional methods of assessing sward height and biomass, species composition, or nutritive value are labor-intensive, yet straightforward. Individuals can be trained in height and biomass data collection and nutritive sample harvesting in short order, and many manuals, tutorials, and courses on taking these measurements are available [57]. However, assessing plant species composition requires knowledge of plant identification or the use of botanical keys despite the simplicity of the processes often used. In contrast, proximal sensing technologies are novel as well as diverse in their equipment, methods, and results. Consequently, potential adopters must first familiarize themselves with the array of options available, learn to use the sensor technology they deem most suitable for their purposes, and then process the data generated by sensors. Alternately, potential adopters may be able to outsource the data processing stage if the service is offered by organizations or companies to which they have access; just such an approach is being tested for remote pasture sensing technology in an American dairy cooperative [58].

5.4. Trialability

In general, the barriers to adoption resulting from the complexity of proximal sensing technologies present similar issues in terms of trialability. For example, if opting to use UAV-mounted proximal sensors, operators must learn to pilot a drone and, depending on local laws, may need certification or licensing to do so [59]. Data processing requires computers with robust processors, as well as software capable of processing raw sensor data [11]. However, some proximal sensing technologies are becoming more accessible and require less expertise to use. Some smartphone apps have emerged for the use of SfM for making 3D maps of objects and scenes. Similarly, a smartphone and tablet were recently released with built-in LiDAR capable of generating point clouds from handheld scanning [60].

Cost is another potential obstacle to the trialability of proximal sensing technologies. Many proximal sensors represent a significant financial investment for land managers such as farmers or conservation professionals, beyond what may constitute an acceptable loss if the technology fails to meet their needs. Again, the emerging LiDAR-equipped tablets and cell phones mentioned above may mitigate cost as an obstacle to adoption, to some extent; the smartphone and tablet are a fraction of the cost of most other proximal sensors and may be used simply by walking through a field or pasture. However, the point clouds generated still require postprocessing for accurate determinations of pasture height. While this may be done with open-source software, it still requires an investment of time to learn the program and process the point cloud.

5.5. Observability

Observability, in Rogers' theory, relates both to the visibility of positive outcomes of an innovation as well as to the visibility of the innovation itself when used by adopters [3]. In terms of observable outcomes, for scientists or natural resources professionals, the volume of data generated by proximal sensing technologies is an observable advantage [46]. For land managers such as farmers, observable advantages of proximal sensing technologies may be more in the spheres of economic or ecological outcomes resulting from the data generated. Prior research demonstrates that innovations that augment the capacity of agricultural producers to make decisions benefiting their economic and ecological goals are more likely to be adopted [56].

6. Conclusions

Proximal sensing technologies are becoming more accessible and useful over time, and their increasing adoption for augmenting data collection programs for pasture and rangeland research appears assured. However, many barriers to the widespread adoption of these technologies by farmers remain: they are often complex to deploy in the field and require postprocessing of data; many proximal sensing systems are expensive, reducing their trialability by producers; and the technologies may require more widespread adoption before sufficient informational and technological support is available. At present, proximal sensors cannot replace traditional fieldwork and on-site monitoring programs, but they can greatly enhance the amount of data collected as well as the scope of data collection possible [11]. We conclude that their adoption will increase as the technology matures and barriers to adoption such as cost and complexity decrease. Additional research quantifying the use of these technologies by agricultural producers is needed, as is qualitative research on perceptions of proximal sensing technologies and barriers to their adoption.

Author Contributions: Conceptualization, B.T.; literature review and writing—original draft preparation, S.G. Both authors have read and agreed to the published version of the manuscript.

Funding: This work is supported by the Cyber-Physical Systems (CPS) (Joint NSF) Competitive Grants Program grant no. 2018-67007-28380 from the USDA National Institute of Food and Agriculture.

Conflicts of Interest: The authors declare no conflict of interest.

References

1. White, R.P.; Murray, S.; Rohweder, M.; Prince, S.D.; Thompson, K.M. *Grassland Ecosystems*; World Resources Institute: Washington, DC, USA, 2000.
2. Bengtsson, J.; Bullock, J.M.; Egoh, B.; Everson, C.; Everson, T.; O'Connor, T.; O'Farrell, P.J.; Smith, H.G.; Lindborg, R. Grasslands—More important for ecosystem services than you might think. *Ecosphere* **2019**, *10*, e02582. [CrossRef]
3. Cislaghi, A.; Giupponi, L.; Tamburini, A.; Giorgi, A.; Bischetti, G.B. The Effects of Mountain Grazing Abandonment on Plant Community, Forage Value and Soil Properties: Observations and Field Measurements in an Alpine Area. *Catena* **2019**, *181*, 104086. [CrossRef]
4. Joyce, L.A.; Briske, D.D.; Brown, J.R.; Polley, H.W.; McCarl, B.A.; Bailey, D.W. Climate Change and North American Rangelands: Assessment of Mitigation and Adaptation Strategies. *Rangel. Ecol. Manag.* **2013**, *66*, 512–528. [CrossRef]
5. Kallenbach, R.L. Describing the Dynamic: Measuring and Assessing the Value of Plants in the Pasture. *Crop. Sci.* **2015**, *55*, 2531–2539. [CrossRef]
6. Svoray, T.; Perevolotsky, A.; Atkinson, P.M. Ecological Sustainability in Rangelands: The Contribution of Remote Sensing. *Int. J. Remote Sens.* **2013**, *34*, 6216–6242. [CrossRef]
7. Rogers, E.M. *Diffusion of Innovations*, 4th ed.; Simon and Schuster: New York, NY, USA, 2010.
8. Mcgrath, C.; Zell, D. The Future of Innovation Diffusion Research and Its Implications for Management: A Conversation with Everett Rogers. *J. Manag. Inq.* **2001**, *10*, 386–391. [CrossRef]
9. Derner, J.D.; Schuman, G.E. Carbon Sequestration and Rangelands: A Synthesis of Land Management and Precipitation Effects. *J. Soil Water Conserv.* **2007**, *62*, 77–85.
10. Holechek, J.L. *An Approach for Setting the Stocking Rate*; Rangel: New York, NY, USA, 1988.
11. Friedel, M.H.; Chewings, V.H.; Bastin, G.N. The Use of Comparative Yield and Dry-Weight-Rank Techniques for Monitoring Arid Rangeland. *Rangel. Ecol. Manag. Range Manag. Arch.* **1988**, *41*, 430–435. [CrossRef]
12. Schalau, J. Rangeland Monitoring: Selecting Key Areas. 2010. Available online: https://extension.arizona.edu/sites/extension.arizona.edu/files/pubs/az1259.pdf (accessed on 12 October 2020).
13. Rayburn, E.B.; Rayburn, S.B. A Standardized Plate Meter for Estimating Pasture Mass in On-Farm Research Trials. *Agron. J.* **1998**, *90*, 238–241. [CrossRef]
14. Dougherty, M.; Burger, J.A.; Feldhake, C.M.; AbdelGadir, A.H. Calibration and Use of Plate Meter Regressions for Pasture Mass Estimation in an Appalachian Silvopasture. *Arch. Agron. Soil Sci.* **2013**, *59*, 305–315. [CrossRef]
15. Bareth, G. Replacing Manual Rising Plate Meter Measurements with Low-Cost UAV-Derived Sward Height Data in Grasslands for Spatial Monitoring. *J. Photogramm. Remote Sens. Geoinf. Sci.* **2018**, *86*, 157–168. [CrossRef]
16. Harmoney, K.R.; Moore, K.J.; George, J.R.; Brummer, E.C.; Russell, J.R. Determination of Pasture Biomass Using Four Indirect Methods. *Agron. J.* **1997**, *89*, 665–672. [CrossRef]
17. Legg, M.; Bradley, S. Ultrasonic Arrays for Remote Sensing of Pasture Biomass. *Remote Sens.* **2020**, *12*, 111. [CrossRef]
18. Moeckel, T.; Safari, H.; Reddersen, B.; Fricke, T.; Wachendorf, M. Fusion of Ultrasonic and Spectral Sensor Data for Improving the Estimation of Biomass in Grasslands with Heterogeneous Sward Structure. *Remote Sens.* **2017**, *9*, 98. [CrossRef]
19. Cunliffe, A.M.; Brazier, R.E.; Anderson, K. Ultra-Fine Grain Landscape-Scale Quantification of Dryland Vegetation Structure with Drone-Acquired Structure-from-Motion Photogrammetry. *Remote Sens. Environ.* **2016**, *183*, 129–143. [CrossRef]
20. Obanawa, H.; Yoshitoshi, R.; Watanabe, N.; Sakanoue, S. Portable LiDAR-Based Method for Improvement of Grass Height Measurement Accuracy: Comparison with SfM Methods. *Sensors* **2020**, *20*, 4809. [CrossRef]
21. Grüner, E.; Astor, T.; Wachendorf, M. Biomass Prediction of Heterogeneous Temperate Grasslands Using an SfM Approach Based on UAV Imaging. *Agronomy* **2019**, *9*, 54. [CrossRef]
22. Rueda-Ayala, V.P.; Peña, J.M.; Höglind, M.; Bengochea-Guevara, J.M.; Andújar, D. Comparing UAV-Based Technologies and RGB-D Reconstruction Methods for Plant Height and Biomass Monitoring on Grass Ley. *Sensors* **2019**, *19*, 535. [CrossRef]
23. Wang, D.; Xin, X.; Shao, Q.; Brolly, M.; Zhu, Z.; Chen, J. Modeling Aboveground Biomass in Hulunber Grassland Ecosystem by Using Unmanned Aerial Vehicle Discrete Lidar. *Sensors* **2017**, *17*, 180. [CrossRef]
24. Pittman, J.J.; Arnall, D.B.; Interrante, S.M.; Moffet, C.A.; Butler, T.J. Estimation of Biomass and Canopy Height in Bermudagrass, Alfalfa, and Wheat Using Ultrasonic, Laser, and Spectral Sensors. *Sensors* **2015**, *15*, 2920–2943. [CrossRef] [PubMed]
25. Legg, M.; Bradley, S. Ultrasonic Proximal Sensing of Pasture Biomass. *Remote Sens.* **2019**, *11*, 2459. [CrossRef]
26. Fricke, T.; Wachendorf, M. Combining Ultrasonic Sward Height and Spectral Signatures to Assess the Biomass of Legume–Grass Swards. *Comput. Electron. Agric.* **2013**, *99*, 236–247. [CrossRef]
27. Schaefer, M.T.; Lamb, D.W. A Combination of Plant NDVI and LiDAR Measurements Improve the Estimation of Pasture Biomass in Tall Fescue (Festuca Arundinacea Var. Fletcher). *Remote Sens.* **2016**, *8*, 109. [CrossRef]
28. Cooper, S.D.; Roy, D.P.; Schaaf, C.B.; Paynter, I. Examination of the Potential of Terrestrial Laser Scanning and Structure-from-Motion Photogrammetry for Rapid Nondestructive Field Measurement of Grass Biomass. *Remote Sens.* **2017**, *9*, 531. [CrossRef]

29. Lu, B.; He, Y. Species Classification Using Unmanned Aerial Vehicle (UAV)-Acquired High Spatial Resolution Imagery in a Heterogeneous Grassland. *ISPRS J. Photogramm. Remote Sens.* **2017**, *128*, 73–85. [CrossRef]
30. Laliberte, A.S.; Rango, A. Image Processing and Classification Procedures for Analysis of Sub-Decimeter Imagery Acquired with an Unmanned Aircraft over Arid Rangelands. *GIScience Remote Sens.* **2011**, *48*, 4–23. [CrossRef]
31. Cruzan, M.B.; Weinstein, B.G.; Grasty, M.R.; Kohrn, B.F.; Hendrickson, E.C.; Arredondo, T.M.; Thompson, P.G. Small Unmanned Aerial Vehicles (Micro-UAVs, Drones) in Plant Ecology. *Appl. Plant. Sci.* **2016**, *4*, 1600041. [CrossRef]
32. Baena, S.; Moat, J.; Whaley, O.; Boyd, D.S. Identifying Species from the Air: UAVs and the Very High Resolution Challenge for Plant Conservation. *PLoS ONE* **2017**, *12*, e0188714. [CrossRef]
33. Miyoshi, G.T.; Arruda, M.d.S.; Osco, L.P.; Marcato Junior, J.; Gonçalves, D.N.; Imai, N.N.; Tommaselli, A.M.G.; Honkavaara, E.; Gonçalves, W.N. A Novel Deep Learning Method to Identify Single Tree Species in UAV-Based Hyperspectral Images. *Remote Sens.* **2020**, *12*, 1294. [CrossRef]
34. Bah, M.D.; Hafiane, A.; Canals, R. Deep Learning with Unsupervised Data Labeling for Weed Detection in Line Crops in UAV Images. *Remote Sens.* **2018**, *10*, 1690. [CrossRef]
35. Alvarez-Taboada, F.; Paredes, C.; Julián-Pelaz, J. Mapping of the Invasive Species Hakea Sericea Using Unmanned Aerial Vehicle (UAV) and WorldView-2 Imagery and an Object-Oriented Approach. *Remote Sens.* **2017**, *9*, 913. [CrossRef]
36. Tay, J.Y.L.; Erfmeier, A.; Kalwij, J.M. Reaching new heights: Can drones replace current methods to study plant population dynamics? *Plant Ecol.* **2018**, *219*, 1139–1150. [CrossRef]
37. Gallacher, D. Drone-Based Vegetation Assessment in Arid Ecosystems. In *Sabkha Ecosystems*; Springer: Berlin, Germany, 2019; pp. 91–98.
38. Gallacher, D.; Khafaga, M.T.; Ahmed, M.T.M.; Shabana, M.H.A. Plant Species Identification via Drone Images in an Arid Shrubland. In Proceedings of the 10th International Rangeland Congress, Saskatoon, SK, Canada, 17–22 July 2016; pp. 981–982.
39. Breckenridge, R.P.; Dakins, M.; Bunting, S.; Harbour, J.L.; Lee, R.D. Using Unmanned Helicopters to Assess Vegetation Cover in Sagebrush Steppe Ecosystems. *Rangel. Ecol. Manag.* **2012**, *65*, 362–370. [CrossRef]
40. Gillan, J.K.; Karl, J.W.; van Leeuwen, W.J. Integrating Drone Imagery with Existing Rangeland Monitoring Programs. *Environ. Monit. Assess.* **2020**, *192*, 1–20. [CrossRef] [PubMed]
41. Laliberte, A.S.; Browning, D.M.; Herrick, J.E.; Gronemeyer, P. Hierarchical Object-Based Classification of Ultra-High-Resolution Digital Mapping Camera (DMC) Imagery for Rangeland Mapping and Assessment. *J. Spat. Sci.* **2010**, *55*, 101–115. [CrossRef]
42. Sun, Y.; Yi, S.; Hou, F. Unmanned Aerial Vehicle Methods Makes Species Composition Monitoring Easier in Grasslands. *Ecol. Indic.* **2018**, *95*, 825–830. [CrossRef]
43. Michez, A.; Philippe, L.; David, K.; Sébastien, C.; Christian, D.; Bindelle, J. Can Low-Cost Unmanned Aerial Systems Describe the Forage Quality Heterogeneity? Insight from a Timothy Pasture Case Study in Southern Belgium. *Remote Sens.* **2020**, *12*, 1650. [CrossRef]
44. Wijesingha, J.; Astor, T.; Schulze-Brüninghoff, D.; Wengert, M.; Wachendorf, M. Predicting Forage Quality of Grasslands Using UAV-Borne Imaging Spectroscopy. *Remote Sens.* **2020**, *12*, 126. [CrossRef]
45. Noland, R.L.; Wells, M.S.; Coulter, J.A.; Tiede, T.; Baker, J.M.; Martinson, K.L.; Sheaffer, C.C. Estimating Alfalfa Yield and Nutritive Value Using Remote Sensing and Air Temperature. *Field Crops Res.* **2018**, *222*, 189–196. [CrossRef]
46. Safari, H.; Fricke, T.; Wachendorf, M. Determination of Fibre and Protein Content in Heterogeneous Pastures Using Field Spectroscopy and Ultrasonic Sward Height Measurements. Comput. *Electron. Agric.* **2016**, *123*, 256–263. [CrossRef]
47. Pullanagari, R.R.; Kereszturi, G.; Yule, I. Integrating Airborne Hyperspectral, Topographic, and Soil Data for Estimating Pasture Quality Using Recursive Feature Elimination with Random Forest Regression. *Remote Sens.* **2018**, *10*, 1117. [CrossRef]
48. Serrano, J.; Shahidian, S.; Marques da Silva, J. Monitoring Seasonal Pasture Quality Degradation in the Mediterranean Montado Ecosystem: Proximal versus Remote Sensing. *Water* **2018**, *10*, 1422. [CrossRef]
49. Duranovich, F.; Yule, I.; Lopez-Villalobos, N.; Shadbolt, N.; Draganova, I.; Morris, S. Using Proximal Hyperspectral Sensing to Predict Herbage Nutritive Value for Dairy Farming. *Agronomy* **2020**, *10*, 1826. [CrossRef]
50. Anderson, K.; Gaston, K.J. Lightweight Unmanned Aerial Vehicles Will Revolutionize Spatial Ecology. *Front. Ecol. Environ.* **2013**, *11*, 138–146. [CrossRef]
51. Pathak, H.S.; Brown, P.; Best, T. A Systematic Literature Review of the Factors Affecting the Precision Agriculture Adoption Process. *Precis. Agric.* **2019**, *20*, 1292–1316. [CrossRef]
52. Kennedy, C.A.; Brunson, M.W. Creating a Culture of Innovation in Ranching. *Rangelands* **2007**, *29*, 35–40. [CrossRef]
53. Michels, M.; von Hobe, C.-F.; Musshoff, O. A Trans-Theoretical Model for the Adoption of Drones by Large-Scale German Farmers. *J. Rural. Stud.* **2020**, *75*, 80–88. [CrossRef]
54. Ghajar, S.; Fernández-Giménez, M.E.; Wilmer, H. Home on the Digital Range: Ranchers' Web Access and Use. *Rangel. Ecol. Manag.* **2019**, *72*, 711–720. [CrossRef]
55. Eastwood, C.; Klerkx, L.; Nettle, R. Dynamics and Distribution of Public and Private Research and Extension Roles for Technological Innovation and Diffusion: Case Studies of the Implementation and Adaptation of Precision Farming Technologies. *J. Rural. Stud.* **2017**, *49*, 1–12. [CrossRef]
56. Bramley, R.G.V.; Ouzman, J. Farmer Attitudes to the Use of Sensors and Automation in Fertilizer Decision-Making: Nitrogen Fertilization in the Australian Grains Sector. *Precis. Agric.* **2019**, *20*, 157–175. [CrossRef]

57. Rangeland Monitoring | Rangelands Gateway. Available online: https://rangelandsgateway.org/topics/maintaining-improving-rangelands/rangeland-monitoring (accessed on 27 July 2021).

58. Organic Valley to Pioneer Use of Satellite Technology to Improve Pasture Grazing. Available online: https://www.organicvalley.coop/newspress/organic-valley-pioneer-use-satellite-technology-improve-pasture-grazing/ (accessed on 24 July 2021).

59. Reger, M.; Bauerdick, J.; Bernhardt, H. Drones in Agriculture: Current and Future Legal Status in Germany, the EU, the USA and Japan. *Landtechnik* **2018**, *73*, 62–80.

60. Perez, S. Snapchat among First to Leverage IPhone 12 Pro's LiDAR Scanner for AR TechCrunch. 2020. Available online: https://social.techcrunch.com/2020/10/13/snapchat-among-first-to-leverage-iphone-12-pros-lidar-scanner-for-ar/ (accessed on 13 October 2020).

Article

Image Generation of Tomato Leaf Disease Identification Based on Adversarial-VAE

Yang Wu and Lihong Xu *

College of Electronics and Information Engineering, Tongji University, Cao'an Road, No. 4800, Shanghai 201804, China; wu_tim@tongji.edu.cn
* Correspondence: xulihong@tongji.edu.cn; Tel.: +86-136-363-62684

Abstract: The deep neural network-based method requires a lot of data for training. Aiming at the problem of a lack of training images in tomato leaf disease identification, an Adversarial-VAE network model for generating images of 10 tomato leaf diseases is proposed, which is used to expand the training set for training an identification model. First, an Adversarial-VAE model is designed to generate tomato leaf disease images. Then, a multi-scale residual learning module is used to replace single-size convolution kernels to enrich extracted features, and a dense connection strategy is integrated into the Adversarial-VAE networks to further enhance the image generation ability. The training set is expanded by the proposed model, which generates the same number of images by training 10,892 images of 10 leaves. The generated images are superior to those of InfoGAN, WAE, VAE, and VAE-GAN measured by the Frechet Inception Distance (FID). The experimental results show that using the extension dataset that is generated by the Adversarial-VAE model to train the Resnet identification model could improve the accuracy of identification effectively. The model proposed in this paper could generate enough images of tomato leaf diseases and provide a feasible solution for data expansion of tomato leaf disease images.

Keywords: Adversarial-VAE; tomato leaf disease identification; image generation; convolutional neural network

Citation: Wu, Y.; Xu, L. Image Generation of Tomato Leaf Disease Identification Based on Adversarial-VAE. *Agriculture* **2021**, *11*, 981. https://doi.org/10.3390/agriculture11100981

Academic Editor: Matt J. Bell

Received: 29 June 2021
Accepted: 6 October 2021
Published: 9 October 2021

Publisher's Note: MDPI stays neutral with regard to jurisdictional claims in published maps and institutional affiliations.

1. Introduction

Leaf disease identification is crucial to control the spread of diseases and advance healthy development of the tomato industry. Well-timed and accurate identification of diseases is the key to early treatment, and an important prerequisite for reducing crop loss and pesticide use. Unlike traditional machine learning classification methods that manually select features, deep neural networks provide an end-to-end pipeline to automatically extract robust features, which significantly improve the availability of leaf identification. In recent years, neural network technology has been widely applied in the field of plant leaf disease identification [1–9], which indicates that deep learning-based approaches have become popular. However, because the deep convolutional neural network (DCNN) has a lot of adjustable parameters, a large amount of labeled data is needed to train the model to improve its generalization ability of the model. Sufficient training images are an important requirement for models based on convolutional neural networks (CNNs) to improve generalization capability. There are little data about agriculture, especially in the field of leaf disease identification. Collecting large numbers of disease data is a waste of manpower and time, and labeling training data requires specialized domain knowledge, which makes the quantity and variety of labeled samples relatively small. Moreover, manual labeling is a very subjective task, and it is difficult to ensure the accuracy of the labeled data. Therefore, the lack of training samples is the main impediment for further improvement of leaf disease identification accuracy. How to train the deep learning model with a small amount of existing labeled data to improve the identification accuracy is a problem worth studying. In general, researchers usually solve this challenge by using traditional data augmentation methods [10]. In computer vision, it makes perfect sense to employ data augmentation, which can change the characteristics of a sample based on prior knowledge so that the newly generated sample also conforms to, or nearly conforms to, the true distribution of the data, while maintaining the sample label. Due to the particularity of image data, additional training data can be obtained from the original image through simple geometric transformation. Common data enhancement methods include rotation, scaling, translation, cropping, noise addition, and so on. However, little additional information can be obtained from these methods.

In recent years, data expansion methods based on generative models have become a research hotspot and have been applied in various fields [11–15]. For example, in [11], the author presents an approach for learning to translate an image from a source domain X to a target domain Y in the absence of paired examples to learn a mapping G: X→Y, such that the distribution of images from G(X) is indistinguishable from the distribution Y using an adversarial loss. Usually, the two most common techniques for training generative models are the generative adversarial network (GAN) [16] and variational auto-encoder (VAE) [17], both of which have advantages and disadvantages. Goodfellow et al. proposed the GAN model [16] for latent representation learning based on unsupervised learning. Through the adversarial learning of the generator and discriminator, fake data consistent with the distribution of real data can be obtained. It can overcome many difficulties, which appear in many tricky probability calculations of maximum likelihood estimation and related strategies. However, because the input z of the generator is a continuous noise signal and there are no constraints, GAN cannot use this z, which is not an interpretable representation. Radford et al. [18] proposed DCGAN, which adds a deep convolutional network based on GAN to generate samples, and uses deep neural networks to extract hidden features and generate data. The model learns the representation from the object to the scene in the generator and discriminator. InfoGAN [19] tried to use z to find an interpretable expression, where z is broken into incompressible noise z and interpretable implicit variable c. In order to make the correlation between x and c, it is necessary to maximize the mutual information. Based on this, the value function of the original GAN model is modified. By constraining the relationship between c and the generated data, c contains interpreted information about the data. In [20], Arjovsky et al. proposed Wasserstein GAN (WGAN), which uses the Wasserstein distance instead of Kullback-Leibler divergence to measure the probability distribution, to solve the problem of gradient disappearance, ensure the diversity of generated samples, and balance sensitive gradient loss between the generator and discriminator. Therefore, WGAN does not need to carefully design the network architecture, and the simplest multi-layer fully connected network can do it. In [17], Kingma et al. proposed a deep learning technique called VAE for learning latent expressions. VAE provides a meaningful lower bound for the log likelihood that is stable during training and during the process of encoding the data into the distribution of the hidden space. However, because the structure of VAE does not clearly learn the goal of generating real samples, it just hopes to generate data that is closest to the real samples, so the generated samples are more ambiguous. In [21], the researchers proposed a new generative model algorithm named WAE, which minimizes the penalty form of the Wasserstein distance between the model distribution and the target distribution, and derives the regularization matrix different from that of VAE. Experiments show that WAE has many characteristics of VAE, and it generates samples of better quality as measured by FID scores at the same time. Dai et al. [22] analyzed the reasons for the poor quality of VAE generation and concluded that although it could learn data manifold, the specific distribution in the manifold it learns is different from the real distribution. In the experiment, it shows that VAE can reconstruct training data well, but it cannot generate new samples well. Therefore, a two-stage VAE is proposed, where the first one is used to learn the position of the manifold, and the second is used to learn the specific distribution within the manifold, which improves the generation effect significantly.

In order to meet the requirements of the training model for the large amount of image data, this paper proposes an image data generation method based on the Adversarial-VAE network model, which expands the image of tomato leaf diseases to generate images of 10 different tomato leaves, overcomes the overfitting problem caused by insufficient training data faced by the identification model. First, the Adversarial-VAE model is designed to generate images of 10 tomato leaves. Then, in view of the obvious differences in the area occupied by the leaves in the dataset and the insufficient accuracy of the feature expression of the diseased leaves using a single-size convolution kernel, the multi-scale residual learning module is used to replace the single-size convolution kernels to enhance the feature extraction ability, and the dense connection strategy is integrated into the Adversarial-VAE model to further enhance the image generative ability. The experimental results show that the tomato leaf disease images generated by Adversarial-VAE have higher quality than InfoGAN, WAE, VAE, and VAE-GAN on the FID. This method provides a solution for data enhancement of tomato leaf disease images and sufficient and high-quality tomato leaf images for different training models, improves the identification accuracy of tomato leaf disease images, and can be used in identifying similar crop leaf diseases.

The rest of the paper is organized as follows: Section 2 introduces the related work. Section 3 introduces the data enhancement methods based on Adversarial-VAE in detail and the detailed structure of the model. In Section 4, the experiment result is described, and the results are analyzed. Finally, Section 5 summarizes the article.

2. Related Work

2.1. Generative Adversarial Network (GAN)

The basic principle of GAN [16] is to obtain the probability distribution of the generator, making the probability distribution of the generator as similar as possible to the probability distribution of the initial dataset, including the generator and discriminator. The generator maps random data to the target probability distribution. In order to simulate the original data distribution as realistically as possible, the target generator should minimize the divergence between the generated data and the real data. Under real conditions, since the data set cannot contain all the information, GAN's generator model cannot fit the probability distribution of the dataset well in practice, and the noise close to the real data is always introduced, so that new information will be generated. In reality, because the dataset cannot contain all the information, the GAN generator model cannot fit the probability distribution of the dataset well in practice, and it will always introduce noise close to the real data, which will generate new information. Therefore, the generated images are allowed to be used as data enhancement for further improving the accuracy of identification. The disadvantage of using GAN to generate images is it uses the random Gaussian noise to generate images, which means that it is not possible to generate any specified type of image. There is no way to decide which random noise can be used to generate the desired image, unless all the initial distribution can be tried. The generator network distinguishes between "real" and "fake" images through a confrontation process. However, the images obtained in this way are only as real as possible, but this does not guarantee that the content of the images is desired. In other words, it is possible that the generator network generates background images to make it as true as possible, but in fact, there is no real target in it.

2.2. Variational Auto-Encoder (VAE)

Variational auto-encoder (VAE) is an important generative model, which was proposed by Diederik P. Kingma and Max Welling [17], including two parts: encoder and decoder.

Figure 1 is the composition model of VAE. The data we can observe is X, and X is generated by the latent variable z; and $z \rightarrow X$ is the generator model from the perspective of the auto-encoder. It is the decoder, and $X \rightarrow z$ is the recognition model, which is similar to the encoder of the auto-encoder. VAE is now widely used to generate images. When the generation model is trained, we can use it to generate images. Unlike GAN, the probability density function (PDF) of the image is known, while GAN does not know the image distribution. Using the auto-encoder can obtain the encoding distribution of such images through the encoding process of the output images, which is equivalent to knowing the corresponding noise distribution to each image, and then the desired image can be generated by selecting specific noise. When generating a new image, you only need to give the model a random implicit vector with a standard normal distribution, so that the desired image can be generated through the decoder, without the need to encode an original image first. In practice, it is necessary to make a trade-off between the accuracy of the model and the factor that the implicit vector obeys the standard normal distribution. The accuracy of the model refers to the degree of similarity between the image generated by the decoder and the original image.

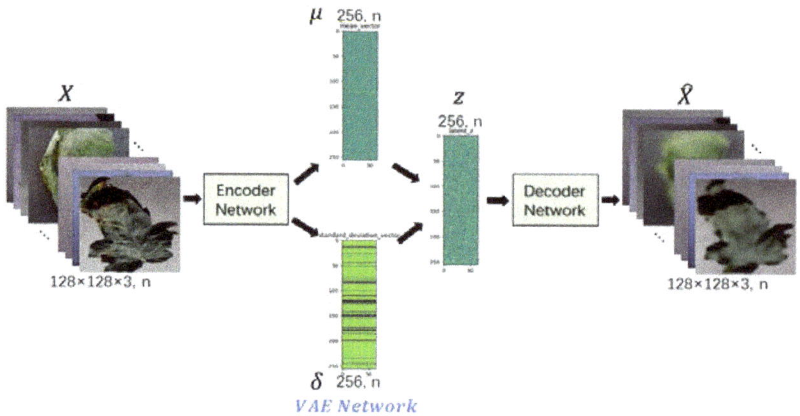

Figure 1. Structure of the VAE network.

2.3. VAE-GAN

VAE-GAN [23] adds a discriminator to the original VAE. If you just operate VAE, the image will be very blurred. After adding the discriminator, the output is forced to be as real as possible. From the perspective of GAN, when training GAN, the generator has never seen what the real image looks like. From the auto-encoder, the generator does not have to cheat the discriminator and has seen what the real image looks like. If you first pass the auto-encoder architecture and the generator has seen a real image, the VAE-GAN will be more stable to learn. VAE-GAN consists of the encoder, generator, and discriminator. The encoder is used to encode, that is, to convert the input image into a vector. The generator is the decoder in VAE, which converts the vector into an output image. Since it is hoped that the output after encoding and decoding is still itself, the input image and output image should be the same as much as possible. The discriminator is used to judge whether the image is realistic or fake (generated by the generator), and gives a scalar (score or probability or binary classification result). The goal of the combination of the encoder and generator is to keep an image as it is after encoding and decoding. Therefore, the updating criterion of the encoder is to minimize the variance of the image before the encoder and after the decoder, and to make the distribution of the image before the encoder and after the decoder as consistent as possible (the distribution is described by KL divergence). The updating criterion of the generator is to minimize the variance of images before the encoder and after the decoder, and the scores of generated and reconstructed images after the discriminator are also as high as possible. The updating criterion of the discriminator is to try to distinguish between the generated, reconstructed, and realistic images, so the scores for the original images are as high as possible, and the scores for the generated and reconstructed images should be as low as possible.

2.4. Two-Stage VAE

VAE is one of the most popular generation models, but the quality of the generation is relatively poor. The gaussian hypothesis of encoders and decoders is generally considered to be one of the reasons for the poor quality of the generation. The authors of [22] carefully analyzed the properties of the VAE objective function, and came to the conclusion that the encoder and decoder gaussian hypothesis of VAE does not affect the global optimal solution. The use of other more complex forms does not obtain a better global optimal solution.

According to [22], VAE can reconstruct training data well but cannot generate new samples well. VAE can learn the manifold where the data is, but the specific distribution in the manifold it learned is different from the real distribution. In other words, every data from the manifold will be perfectly reconstructed after VAE. For this reason, the first VAE is used to learn the position of the manifold, and the second VAE is used to learn the specific distribution within the manifold. Specifically, the first VAE transforms the training data into a certain distribution in the hidden space, which occupies the entire hidden space instead of on the low-dimensional manifold. The second VAE is used to learn the distribution in the hidden space since the latent variable occupies the entire hidden space dimension. Therefore, according to the theory, the second VAE can learn the distribution in the hidden space of the first VAE.

3. Materials and Methods

3.1. Dataset

PlantVillage [24] is an internet public image library of plant leaf diseases initiated and established by David, an epidemiologist at the University of Pennsylvania. This dataset collects more than 50,000 images of 14 species of plants with 38 category labels. Among them, 18,162 tomato leaves of 10 categories, which are respectively healthy leaves and 9 kinds of diseased leaves, were used as the basic data set of crop disease images for the experiment. Figure 2 shows an example of 10 tomato leaves. In the practical application, the image size was changed to 128×128 pixels during preprocessing in order to reduce both the calculation and training time of model.

Figure 2. Examples of tomato leaf diseases: healthy, Tomato bacterial spot (TBS), Tomato early blight (TEB), Tomato late blight (TLB), Tomato leaf mold (TLM), Tomato mosaic virus (TMV), Tomato septoria leaf spot (TSLS), Tomato target spot (TTS), Tomato two-spotted spider mite (TTSSM), and Tomato yellow leaf curl virus (TYLCV), respectively.

3.2. Adversarial-VAE Model for Generating Tomato Leaf Disease Images

The deep neural network has a large number of adjustable parameters, so it needs a large amount of labeled data to improve the generalization ability of the model. However, there has always been a data vacuum in agriculture, making it difficult to collect a lot of data. At the same time, it is also difficult to label all collected data accurately. Due to a lack of experience, it is difficult to judge whether the identification is accurate, so experienced experts are needed to accurately label the data. In order to meet the requirements of the training model for the large amount of image data, this paper proposes an image data generation method based on the Adversarial-VAE network model, which expands the tomato leaf disease images in the PlantVillage dataset, and overcomes the problem of over-fitting caused by insufficient training data faced by the identification model.

3.2.1. Adversarial-VAE Model

The Adversarial-VAE model of tomato leaf disease images consists of stage 1 and stage 2. Stage 1 is a VAE-GAN network, consisting of an encoder (E), generator (G), and discriminator (D). Stage 2 is a VAE network, consisting of an encoder (E) and decoder (D).

The detailed model of Adversarial-VAE is shown in Figure 3. In stage 1, the input images are encoded and decoded, and the discriminator is used to determine whether the images are real or fake to improve the model's generation ability. The input to the model is an image X of size $128 \times 128 \times 3$, which is compressed into two vectors μ and σ with a size of 256 after passing through the encoder network, and then combined into a latent vector z with a size of 256. After passing through the generator network, size expansion is realized to generate an image $\overline{X}$ with a size of $128 \times 128 \times 3$. The input of the discriminator network is the original image X, generated image $\hat{X}$, and reconstructed image $\overline{X}$ to determine whether the image is real or fake. Stage 2 encodes and decodes the latent variable z. Specifically, stage 1 transforms the training data X into some distribution z in the latent space, which occupies the whole latent space rather than on the low-dimensional manifold of the latent space. Stage 2 is used to learn the distribution in the latent space. Since latent variables occupy the whole dimension, according to the theory [22], stage 2 can learn the distribution in the latent space of stage 1. After the Adversarial-VAE model is trained, z is sampled from the gaussian model and $\overline{z}$ is obtained through stage 2. $\overline{z}$ is obtained through the generator network of stage 1 to obtain $\hat{X}$, which is the generated sample and is used to expand the training set in the subsequent identification model.

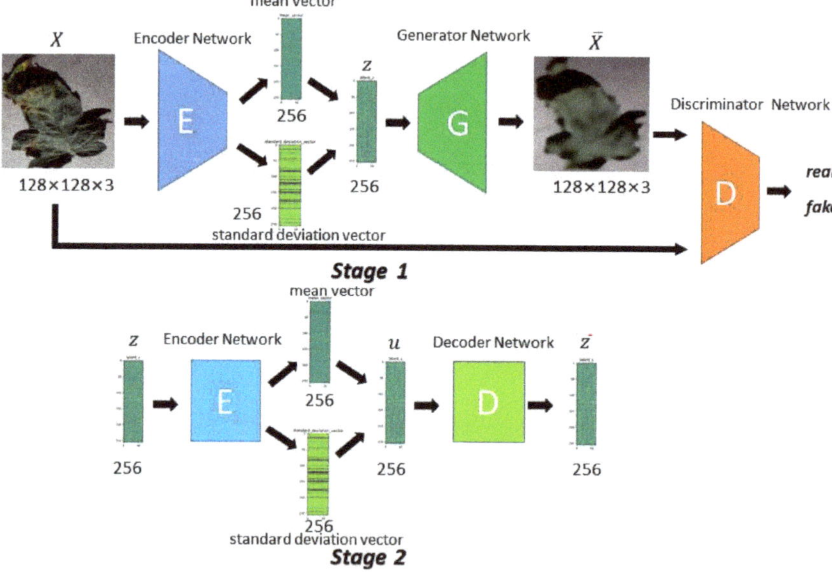

Figure 3. Structure of the Adversarial-VAE model.

3.2.2. Components of Stage 1

Stage 1 is a VAE-GAN network composed of an encoder (E), generator (G), and discriminator (D). It is used to transform training data into a certain distribution in the hidden space, which occupies the entire hidden space rather than on the low-dimensional manifold. The encoder converts an input image X of size $128 \times 128 \times 3$ into two vectors of mean and variance of size 256. The detailed encoder network of stage 1 is shown in Figure 4 and the output sizes of every layer are shown in Table 1. The encoder network consists of a series of convolution layers. It is composed of Conv, 4 layers, Scale, Reducemean, Scale_fc and FC. The 4 layers is made up of four alternating Scale and Downsample, and Scale is the ResNet module, which is used to extract features. Downsample is used to decrease the size of each feather map and increase the number of channels. After each layer, the number of channels is doubled and the size is halved. The input of the model is a $128 \times 128 \times 3$ image, the size of the input vector is changed to $128 \times 128 \times 16$ after Conv layer, while after 4 layers, the size is $8 \times 8 \times 256$. Reducemean is global pooling, and the structure of Scale_fc is shown in Figure 4 for better access to global information.

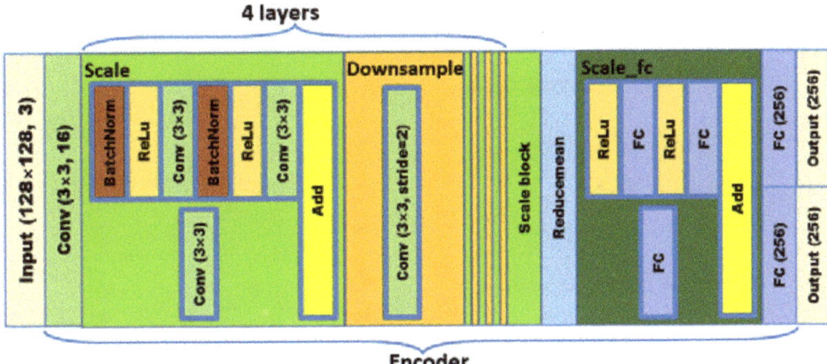

Figure 4. Encoder network.

Table 1. Output size of the layer in the encoder network.

Layer	Input	Conv	Scale 0	Downsample 0	Scale 1	Downsample 1
Size	$128 \times 128 \times 3$	$128 \times 128 \times 16$	$128 \times 128 \times 16$	$64 \times 64 \times 32$	$64 \times 64 \times 32$	$32 \times 32 \times 64$
Layer	... ...	Downsample 3	Scale 4	Reducemean	Scale_fc	FC
Size	... ...	$8 \times 8 \times 256$	$8 \times 8 \times 256$	256	256	256

The generator is both VAE's decoder and GAN's generator, and they have the same function: converting vector to $\overline{X}$. The decoder is used to decode, restoring the latent vector z of size 256 to an image of size $128 \times 128 \times 3$. The goal of the combination of the encoder and generator is to keep an image as original as possible after the encoder and generator. The detailed generator network of stage 1 is shown in Figure 5 and related parameters are shown in Table 2. The generator network consists of a series of deconvolution layers, which is composed of FC, 6 layers, and Conv. FC means fully connected. The input of the model is a vector with 256, which is drawn from a gaussian distribution or reparameterization from the output of the encoder network. The size is changed to 4096 after FC and to $2 \times 2 \times 1024$ after Reshape further. Six layers are made up of six alternating Upsample and Scale. Upsample is deconvolution layer, which is used to expand the size of the feature map and reduce the number of channels. After each Upsample, the length and width of the feature map are doubled, and the number of channels is halved. Scale is the Resnet module, which is used to extract features. After 6 layers, the size is changed to $128 \times 128 \times 3$.

Additionally, after Conv, the size is changed to $128 \times 128 \times 3$, which is the same size as the input image.

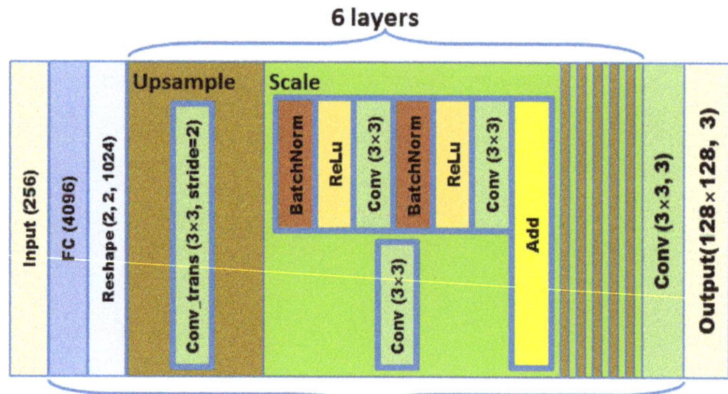

Figure 5. Generator network.

Table 2. Output size of the layer in the generator network.

Layer	Input	FC	Reshape	Upsample 0	Scale 0	Upsample 1
Size	256	4096	$2 \times 2 \times 1024$	$4 \times 4 \times 512$	$4 \times 4 \times 512$	$8 \times 8 \times 256$
Layer	… …	**Upsample 4**	**Scale 4**	**Upsample 5**	**Scale 5**	**Conv**
Size	… …	$64 \times 64 \times 32$	$64 \times 64 \times 32$	$128 \times 128 \times 16$	$128 \times 128 \times 16$	$128 \times 128 \times 3$

The discriminator will be able to differentiate the generated, reconstructed, and realistic images as much as possible. Therefore, the score for the original image should be as high as possible, and the scores for the generated and reconstructed images should be as low as possible. Its structure is similar to that of the encoder, except that the final two FCs with a size of 256 are generated at the end and replaced with FC with a size of 1. The output is true or false, which is used to enhance the image generation ability of the network, making the generated image more like the real image. The details are shown in Figure 6 and related parameters are shown in Table 3.

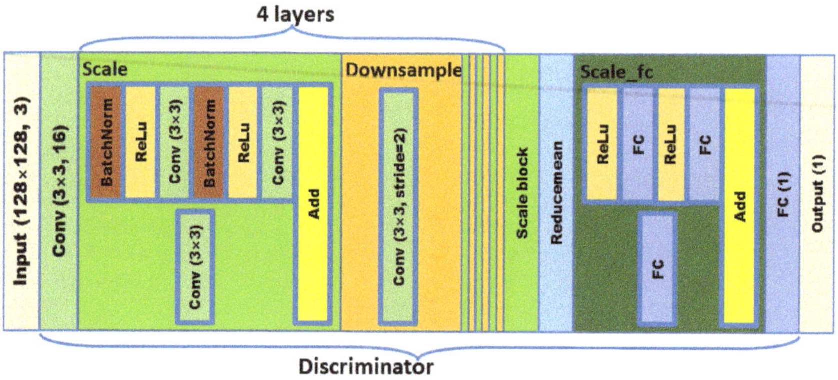

Figure 6. Discriminator network.

Table 3. Output size of the layer in the discriminator network.

Layer	Input	Conv	Scale 0	Downsample 0	Scale 1	Downsample 1
Size	$128 \times 128 \times 3$	$128 \times 128 \times 16$	$128 \times 128 \times 16$	$64 \times 64 \times 32$	$64 \times 64 \times 32$	$32 \times 32 \times 64$
Layer	... ...	**Downsample 3**	**Scale 4**	**Reducemean**	**Scale_fc**	**FC**
Size	... ...	$8 \times 8 \times 256$	$8 \times 8 \times 256$	256	256	1

3.2.3. Components of Stage 2

Stage 2 is a VAE network consisting of the encoder (E) and decoder (D), which is used to learn the distribution of hidden space in stage 1 since the latent variables occupy the entire latent space dimension. Both the encoder (E) and decoder (D) are composed of a fully connected layer. The structure is shown in Figure 7. The input of the model is a latent vector with size 256, which is drawn from a gaussian distribution.

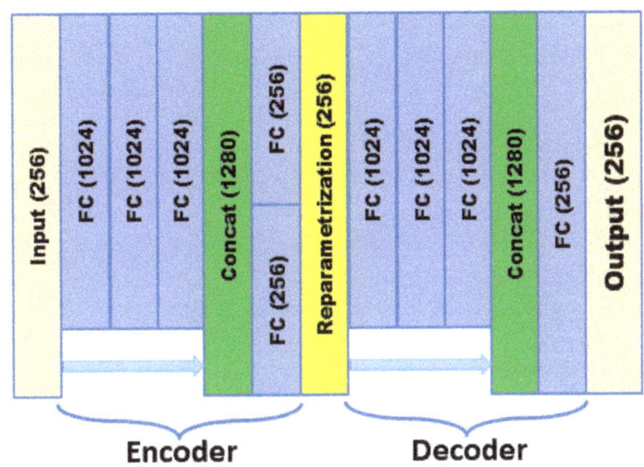

Figure 7. Structure of stage 2 in the Adversarial-VAE model.

3.3. Improved Adversarial-VAE Model

3.3.1. Multi-Scale Convolution

In the PlantVillage dataset, there are obvious differences in the area occupied by the leaves in the image, so the single-size convolution kernel is not accurate enough to check the feature expression of disease leaves. Therefore, in order to make the extracted features more abundant, a multi-scale convolution kernel is applied instead of a single-size convolution kernel to construct the residual learning module so that tomato disease identification can achieve a higher accuracy rate, as shown in Figure 8.

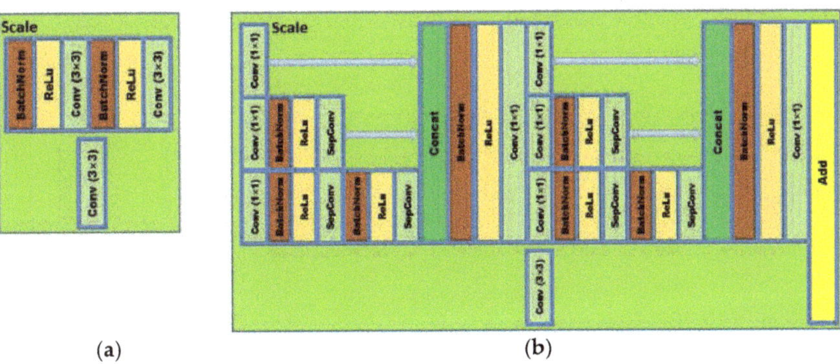

Figure 8. The original structure (**a**) and improved structure (**b**) of the Scale module.

In order to utilize the multi-scale convolution kernel, the convolutional layer in the original residual learning module designed according to the Inception [25] structure, and the computational amount required for the 5×5 convolution kernel is relatively large to reduce the number of parameters and increase the calculation speed. During practical application, the 5×5 convolution kernel is replaced by two 3×3 convolution kernels, which does not allow the convolution layer to be extracted to different levels with different receptive fields.

Specifically, a single 3×3 convolution kernel (Conv (3×3)) in ResNet is replaced by multiple convolution kernels to expand the convolution width, and the information obtained from each convolution kernel is added up through Concat. After BatchNorm and Relu, the mixed feature of Conv (1×1) is used as the input of the next operation. Multiple convolution cores here refer to 1×1 convolution kernel (Conv (1×1)), 1×1 convolution (Conv (1×1)) followed by separable convolution (SepConv), and 1×1 convolution (Conv (1×1)) followed by separable convolution (SepConv) followed by separable convolution (SepConv). Depthwise convolutions are also used to construct a lightweight deep neural network. In this case, the standard convolution is decomposed into depthwise convolution and pointwise convolution. Each channel is convolution individually, which is used to combine the information of each channel to reduce model parameters and computation.

3.3.2. Dense Connection Strategy

As another CNN with a deeper number of layers, Densenet has fewer parameters than Resnet. Its bypass enhances the reuse of features, and the network is easier to train and has a certain regularization effect, and alleviates the problems of gradient vanishing and model degradation. The problem of gradient disappearance is more likely to occur when the network depth is deeper. The reason is that the input information and gradient information are transmitted between many layers. Now, dense connection is equivalent to each layer directly connecting input and loss, so the phenomenon of gradient disappearance can be reduced and the network depth can be increased. Therefore, the dense connection strategy from DenseNet [26] is applied to the encoder network and generator network in stage 1. Each layer uses the feature map as the input of the latter layer, which can effectively extract the features of the lesion and alleviate the disappearing gradient. As shown in Figure 9, due to the inconsistency of the feature scales of the front and back layers, 1×1 convolution is used to achieve the consistency of feature scales. The dense connection strategy shares the weights of the prior layers and improves the feature extraction capabilities.

3.4. Loss Function

Stage 1 is VAE-GAN network. In stage 1, the goal of the encoder and generator is to keep an image as original as possible after code. The goal of the discriminator is to try to differentiate the generated, reconstructed, and realistic images. The training pipeline of the stage 1 Algorithm 1 is as follows:

Algorithm 1: The training pipeline of the stage 1.

Initial parameters of the models: θ_e, θ_g, θ_d
while training **do**
    $x^{real} \leftarrow$ batch of images sampled from the dataset.
    $z_\mu^{real}, z_\sigma^{real} \leftarrow E_{\theta_e}(x^{real})$
    $z^{real} \leftarrow z_\mu^{real} + \varepsilon z_\sigma^{real}$ with $\varepsilon \sim N(0, Id)$
    $\overline{x}^{real} \leftarrow G_{\theta_g}(z^{real})$
    $z^{fake} \leftarrow$ prior $P(z)$
    $x^{fake} \leftarrow G_{\theta_g}(z^{fake})$
{Compute losses gradients and update parameters.}
$$\theta_e \leftarrow \|\overline{x}^{real} - x^{real}\| + KL(P(z^{real}|x^{real})\|P(z))$$
$$\theta_g \leftarrow \|\overline{x}^{real} - x^{real}\| - D_{\theta_d}(\overline{x}^{real}) - D_{\theta_d}(x^{fake})$$
$$\theta_d \leftarrow D_{\theta_d}(\overline{x}^{real}) + D_{\theta_d}(x^{fake}) - D_{\theta_d}(x^{real})$$
end while

Stage 2 is the VAE network. In stage 2, the goal of the encoder and decoder is to keep an image as original as possible after codec. Therefore, the updating criterion of the encoder is to minimize the variance of the image before the encoder and after the decoder, and to make the distribution of the image as consistent as possible before the encoder and after the decoder. The updated criterion of the decoder is to minimize the variance of images before the encoder and after the decoder. The training pipeline of the stage 2 Algorithm 2 is as shown below:

Algorithm 2: The training pipeline of the stage 2.

Initial parameters of the models: θ_e, θ_d.

while training **do**

$\quad z^{real} \leftarrow$ Gaussian distribution.

$\quad u_\mu{}^{real}, u_\sigma{}^{real} \leftarrow E_{\theta_e}(z^{real})$.

$\quad u^{real} \leftarrow u_\mu{}^{real} + \varepsilon u_\sigma{}^{real}$ with $\varepsilon \sim N(0, Id)$.

$\quad \bar{z}^{real} \leftarrow D_{\theta_d}(u^{real})$.

$\quad u^{fake} \leftarrow$ prior $P(u)$.

$\quad z^{fake} \leftarrow D_{\theta_d}(u^{fake})$.

{Compute losses gradients and update parameters.}

$\quad \theta_e \Leftarrow \|\bar{z}^{real} - z^{real}\| + KL(P(u^{real}|z^{real}) \| P(u))$.

$\quad \theta_d \Leftarrow \|\bar{z}^{real} - z^{real}\|$.

end while

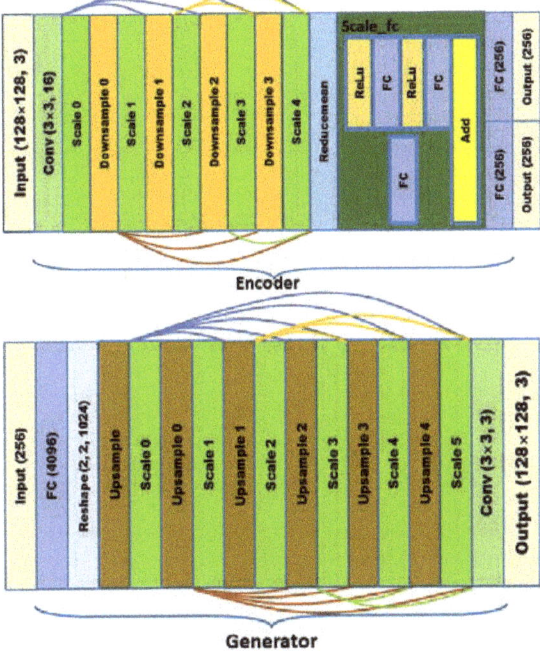

Figure 9. Dense connection strategy in the encoder and generator.

3.5. Experimental Setup

The experimental configuration environment of this paper is as follows: Ubuntu16.04 LST 64-bit system, processor Intel Core i5-8400 (2.80 GHz), memory is 8 GB, graphics card is GeForce GTX1060 (6G), and using the Tensorflow-GPU1.4 deep learning framework with python programming language.

3.6. Performance Evaluation Metrics

The FID evaluation model is introduced to evaluate the performance of the image generation task. The FID score was proposed by Martin Heusel [27] in 2017. It is a metric for evaluating the quality of the generated image and is specifically used to evaluate the performance of GAN. It is a measure of the distance between the feature vector of the real image and the generated image. This score is proposed as an improvement on the existing inception score (IS) [28,29]. It calculates the similarity of the generated image to the real image, which is better than the IS. The disadvantage of IS is that it does not use statistics from the true sample and compare them to statistics from the generated sample.

As with the IS, the FID score uses the Inception V3 model. Specifically, the coding layer of the model (the last pooled layer before the classified output of the image) is used to extract the features specified by computer vision techniques for the input image. These activation functions are calculated for a set of real and generated images. By calculating the mean value and covariance of the image, the output of the activation function is reduced to a multivariable gaussian distribution. These statistics are then used to calculate the real image and generate activation functions in the image collection. The FID is then used to calculate the distance between the two distributions. The lower the FID score, the better the image quality. On the contrary, the higher the score, the worse the quality.

4. Results and Discussion

In order to verify the effectiveness of the leaf disease identification model proposed in this paper, a total of 18,162 images of the tomato disease from PlantVillage are randomly divided into a training set, verification set, and test set, of which the training set accounts for about 60%, which means 10,892 images, as shown in Table 4. The verification set accounts for about 20% or 3632 images, and the test set accounts for about 20% or 3636 images. They are used to train the model, select the model, and evaluate the performance of the proposed model.

Table 4. Detailed information of the tomato leaf disease dataset.

Class	All Sample Numbers	60% of Sample Numbers
healthy	1592	954
TBS	2127	1276
TEB	1000	600
TLB	1910	1145
TLM	952	571
TMV	373	223
TSLS	1771	1062
TTS	1404	842
TTSSM	1676	1005
TYLCV	5357	3214
ALL	18,162	10,892

The Adversarial-VAE model is used to generate training samples, and the number of generated samples is consistent with the number of samples corresponding to the original training set, so the sample size is doubled, and the generated data is added to the training set. For these datasets with generated images, all the generated images are placed in the training set, and all the images in the test set are from the initial dataset. The test set is completely derived from the initial dataset. The flowchart of the data augmentation method is shown in Figure 10. In the figure, generative model refers to the generation part of the Adversarial-VAE model, which is composed of stage 2 and the generator network in stage 1. After the Adversarial-VAE model is trained, z is sampled from the Gaussian model, and $\bar{z}$ is obtained through stage 2, and $\overline{X}$ is obtained through the generator network of stage 1, which is the generated sample. For 10 kinds of tomato leaf images, we train 10 Adversarial-VAE models. For each class, we generate samples by sampling vectors corresponding to the number of categories from the gaussian model in order to generate a different number of samples.

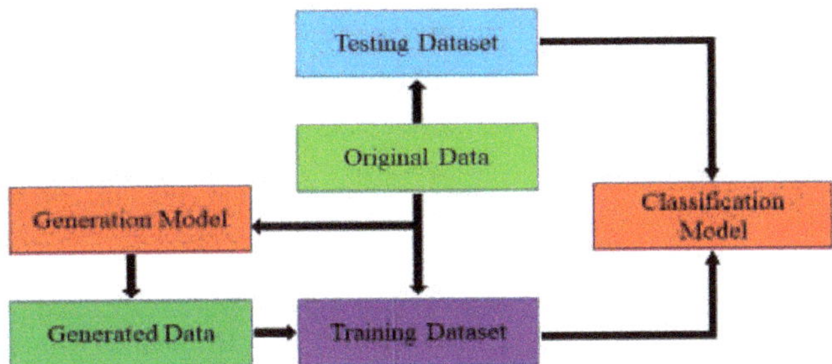

Figure 10. The workflow of the image generation based on Adversarial-VAE networks.

4.1. Generation Results and Analysis

The proposed Adversarial-VAE networks are compared with several advanced generation methods, including InfoGAN, WAE, VAE, VAE-GAN, and 2VAE, which are used to generate tomato diseased leaf images. We compare the reconstructed image quality and the generated image quality through the FID score as shown in Tables 5 and 6. Table 5 lists the FID of the reconstruction images under the different neural network models. Table 6 shows the FID comparison between different generative methods. Reconstruction-FID demonstrates the ability of this method to reconstruct the original input image. The lower the value is, the better the reconstruction capability is. Generation-FID demonstrates the ability of this method to generate new images. The lower the value is, the better the reconstruction capability is.

Tables 5 and 6 show Reconstruction-FID and Generation-FID of 10 kinds of tomato leaf images, respectively. From the tables, we can see that WAE is better at reconstruction of the images than other methods. The average FID score is 105.74, which is the lowest score, and it also obtained the lowest score in most categories except TBS and TYLCV, which means WAE has excellent ability in reconstruction. Adversarial-VAE is the best in the generation of the images. The average FID score is 161.77, which is the lowest score, and it also obtained the lowest score in most categories, which means Adversarial-VAE has more advantages in generation than the others.

Table 5. Reconstruction-FID comparison between different generative methods.

Reconstruction-FID	InfoGAN [19]	WAE [21]	VAE [17]	VAE-GAN [23]	2VAE [22]	Adversarial-VAE
healthy	172.61	129.47	155.64	130.08	155.64	130.08
TBS	135.29	103.11	148.07	114.24	148.07	114.24
TEB	126.96	106.69	138.87	100.59	138.87	100.59
TLB	180.10	111.81	169.80	119.23	169.80	119.23
TLM	160.93	133.79	161.37	147.08	161.37	147.08
TMV	144.71	125.86	157.20	140.23	157.20	140.23
TSLS	120.24	90.43	139.41	108.57	139.41	108.57
TTS	107.88	81.74	137.89	99.67	137.89	99.67
TTSSM	114.22	91.23	141.42	106.89	141.42	106.89
TYLCV	140.11	83.23	133.05	79.76	133.05	79.76
AVERAGE	140.31	105.74	148.27	114.63	148.27	114.63

Generation-FID of Adversarial-VAE alone, Adversarial-VAE + multi-scale convolution, Adversarial-VAE + dense connection strategy, and the improved Adversarial-VAE, which used multi-scale convolution and the dense connection strategy, are compared in Table 7. The average FID score is 156.96, which is the lowest score, and it also obtained the lowest score in most categories. As can be seen from the table, the improved model reduced the FID score for most types of disease, with an average FID score reduction of 4.81. It shows that the improved model has a better generative ability. The generated images are shown in Figure 11 based on Adversarial-VAE. And Figure 12 shows the generated images based on VAE networks.

Table 6. Generation-FID comparison between different generative methods.

Generation-FID	InfoGAN [19]	WAE [21]	VAE [17]	VAE-GAN [23]	2VAE [22]	Adversarial-VAE
healthy	221.86	202.06	186.37	167.46	179.83	162.57
TBS	232.88	221.85	190.71	178.75	187.09	179.96
TEB	183.09	169.42	158.43	132.42	153.65	133.65
TLB	277.65	227.51	192.38	184.64	199.17	180.71
TLM	235.07	219.42	200.15	200.90	196.47	197.45
TMV	210.91	211.38	191.24	214.60	196.78	210.54
TSLS	199.31	182.59	156.61	148.31	152.93	146.11
TTS	199.87	208.23	191.90	163.99	185.01	161.07
TTSSM	195.08	210.70	175.97	147.95	173.95	146.83
TYLCV	182.74	172.82	151.22	99.60	146.89	98.76
AVERAGE	213.85	202.60	179.50	163.86	177.18	161.77

Table 7. Generation-FID comparison of the proposed generative method.

Generation-FID	Adversarial-VAE Alone	Adversarial-VAE + Multi-Scale Convolution	Adversarial-VAE + Dense Connection Strategy	Improved Adversarial-VAE
healthy	162.57	162.64	167.63	171.63
TBS	179.96	170.29	176.3	167.53
TEB	133.65	128.28	130.81	126.84
TLB	180.71	175.15	170.42	166.92
TLM	197.45	194.81	191.42	187.79
TMV	210.54	202.39	198.28	189.09
TSLS	146.11	151.91	147.11	151.8
TTS	161.07	155.89	166.72	165.84
TTSSM	146.83	144.54	143.74	142.32
TYLCV	98.76	98.31	98.64	99.79
AVERAGE	161.77	158.42	159.11	156.96

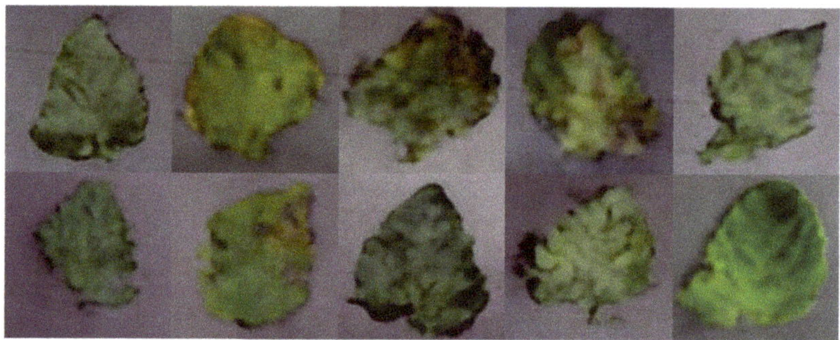

Figure 11. Examples of tomato diseased leaf generated by improved Adversarial-VAE networks: healthy, TBS, TEB, TLB, TLM, TMV, TSLS, TTS, TTSSM, and TYLCV, respectively.

Figure 12. Examples of tomato diseased leaf generated by VAE networks: healthy, TBS, TEB, TLB, TLM, TMV, TSLS, TTS, TTSSM, and TYLCV, respectively.

4.2. Identification Results and Analysis

The original training set contains 10,892 images. After using improved Adversarial-VAE, the training set is expanded to 21,784 images. For comparative experiments, the original data set is expanded twice by replication, namely 21,784 images. Three experiments are carried out to train the classification network as shown in Figure 13 to identify tomato leaf diseases. During the operation, the training set and the test set are divided into batches by batch training. The batch training method is used to divide the training set and the test set into multiple batches. Each batch trains 32 images, that is, the minibatch is set to 32. After training 4096 images, the verification set is used to determine the retained model. After training all the training set images, the test set is tested. Each test batch is set to 32. All the images in a training set are iterated through as an iteration (epoch) for a total of 10 iterations. The model is optimized in using the momentum optimization algorithm and the learning rate is set at 0.001.

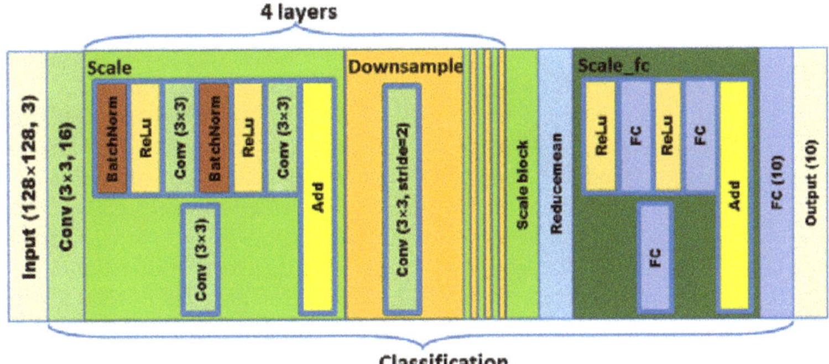

Figure 13. Structure of the classification network.

Table 8 shows the classification accuracy of the classification network trained with the expanded training set generated by different generative methods. After training the classification network with the original training set, the identification accuracy on the test set is 82.87%; With the double original training set, the identification accuracy on the test set is 82.95%, and after training the classification network with the training set expanded by improved Adversarial-VAE, the identification accuracy reaches 88.43%, an increase of 5.56%. Compared with the double original training set, it also improved by 5.48%, which proves the effectiveness of the data expansion. The InfoGAN and WAE generative models were used to generate samples for the training the classification network, but the classification accuracy was not improved, which can be understood as poor sample generation and no effect was mentioned for training, as shown in Table 8.

Table 8. Classification accuracy of the classification network trained with the expanded training set generated by different generative methods.

	Classification Alone	InfoGAN + Classification	WAE + Classification	VAE + Classification	VAE-GAN + Classification	2VAE + Classification	Improved Adversarial-VAE + Classification
Accuracy	82.87%	82.42%	82.16%	84.65%	86.86%	85.43%	88.43%

5. Conclusions

Leaf disease identification is the key to control the spread of disease and ensure healthy development of the tomato industry. The deep neural network-based method requires a lot of data for training. However, there is little data in many agricultural fields. In the field of tomato leaf disease identification, it is a waste of manpower and time to collect large-scale labeled data. Labeling of training data requires very professional knowledge. All these factors lead to either the number and category of labeling being relatively small, or the labeling data for a certain category being very small, and manual labeling is very subjective work, which makes it difficult to ensure high accuracy of the labeled data.

To solve the problem of a lack of training images of tomato leaf diseases, an Adversarial-VAE network model was proposed to generate images of 10 different tomato leaf diseases to train the recognition model. Firstly, an Adversarial-VAE model was designed to generate tomato leaf disease images. Then, the multi-scale residuals learning module was used to replace the single-size convolution kernel to enhance the ability of feature extraction, and the dense connection strategy was integrated into the Adversarial-VAE model to further enhance the ability of image generation. The Adversarial-VAE model was only used to generate training data for the recognition model. During the training and testing phase of the recognition model, no computation and storage costs were introduced in the actual model deployment and production environment. A total of 10,892 tomato leaf disease images were used in the Adversarial-VAE model, and 21,784 tomato leaf disease images were finally generated. The image of tomato leaf diseases based on the Adversarial-VAE model was superior to the InfoGAN, WAE, VAE, and VAE-GAN methods in FID. The experimental results show that the proposed Adversarial-VAE model can generate enough of the tomato plant disease image, and image data for tomato leaf disease extension provides a feasible solution. Using the Adversarial-VAE extension data sets is better than using other data expansion methods, and it can effectively improve the identification accuracy, and can be generalized in identifying similar crop leaf diseases. In future work, in order to improve the robustness and accuracy of identification, we will continue to find better data enhancement methods to solve the problem of tomato leaf disease identification, which can be applied to the detection networks.

This method was proposed based on improving the classification accuracy on the basis of many labeled samples. At the beginning of this study, the most direct way was to expand each class with one network, so that when new categories need to be added, only one network needs to be trained with the samples of the new category, instead of retraining with all samples. We also considered training only one network to generate data samples of different categories by adding an input as a category control, but this has the side effect of requiring several networks to be retrained if new categories need to be generated. If there is no large amount of annotated data as training samples for training the generative model, for example, disease leaves for another plant cannot cover the sample space, the generative model cannot be directly trained in this way, and the number of samples needs to be expanded first. In practice, it is difficult to collect disease leaf images, so the problem of few-shot learning needs to be solved urgently. In summary, we will strive to achieve continuous improvement of the performance and try to apply it to practical agricultural production.

Author Contributions: All authors provided ideas of the proposed method and amended the manuscript; Y.W. designed the experiments and organized the experimental data. L.X. established the guidance for the research idea, authored or reviewed drafts of the paper, approved the final draft. Both authors have read and agreed to the published version of the manuscript.

Funding: This research was funded by Shanghai Agriculture Applied Technology Development Program of China (Grant No. 2019-02-08-00-07-F01121), National Natural Science Foundation of China (Grant No. 61973337), US National Science Foundation's BEACON Center for the Study of Evolution in Action (#DBI-0939454).

Institutional Review Board Statement: Not applicable.

Informed Consent Statement: Not applicable.

Data Availability Statement: The datasets generated for this study are available on request to the corresponding author.

Conflicts of Interest: The authors declare no conflict of interest.

References

1. Stewart, E.L.; Wiesner-Hanks, T.; Kaczmar, N.; DeChant, C.; Wu, H.; Lipson, H.; Nelson, R.J.; Gore, M.A. Quantitative Phenotyping of Northern Leaf Blight in UAV Images Using Deep Learning. *Remote Sens.* **2019**, *11*, 2209. [CrossRef]
2. Zhang, Y.; Song, C.; Zhang, D. Deep learning-based object detection improvement for tomato disease. *IEEE Access* **2020**, *8*, 56607–56614. [CrossRef]
3. Xie, X.; Ma, Y.; Liu, B.; He, J.; Li, S.; Wang, H. A Deep-Learning-Based Real-Time Detector for Grape Leaf Diseases Using Improved Convolutional Neural Networks. *Front. Plant Sci.* **2020**, *11*, 751. [CrossRef] [PubMed]
4. Saleem, M.H.; Potgieter, J.; Arif, K.M. Plant Disease Detection and Classification by Deep Learning. *Plants* **2019**, *11*, 468. [CrossRef] [PubMed]
5. Afzaal, H.; Farooque, A.A.; Schumann, A.W.; Hussain, N.; McKenzie-Gopsill, A.; Esau, T.; Abbas, F.; Acharya, B. Detection of a Potato Disease (Early Blight) Using Artificial Intelligence. *Remote Sens.* **2021**, *13*, 411. [CrossRef]
6. Wu, Y.; Xu, L. Crop Organ Segmentation and Disease Identification Based on Weakly Supervised Deep Neural Network. *Agronomy* **2019**, *9*, 737. [CrossRef]
7. Tetila, E.C.; Machado, B.B.; Menezes, G.K.; Oliveira, A.D.S.; Alvarez, M.; Amorim, W.P.; De Souza Belete, N.A.; Goncalves Da Silva, G.; Pistori, H. Automatic recognition of soybean leaf diseases using UAV images and deep convolutional neural networks. *IEEE Geosci Remote Sens Lett.* **2019**, *17*, 903–907. [CrossRef]
8. Yu, H.J.; Son, C.H. Leaf spot attention network for apple leaf disease identification. In Proceedings of the IEEE/CVF Conference on Computer Vision and Pattern Recognition Workshops, Seattle, WA, USA, 14–19 June 2020; pp. 52–53.
9. Liu, X.; Min, W.; Mei, S.; Wang, L.; Jiang, S. Plant Disease Recognition: A Large-Scale Benchmark Dataset and a Visual Region and Loss Reweighting Approach. *IEEE Trans. Image Process.* **2021**, *30*, 2003–2015. [CrossRef] [PubMed]
10. Jiang, P.; Chen, Y.; Liu, B.; He, D.; Liang, C. Real-Time Detection of Apple Leaf Diseases Using Deep Learning Approach Based on Improved Convolutional Neural Networks. *IEEE Access* **2019**, *7*, 59069–59080. [CrossRef]
11. Zhu, J.; Park, T.; Isola, P.; Efros, A. Unpaired Image-To-Image Translation Using Cycle-Consistent Adversarial Networks. In Proceedings of the IEEE International Conference on Computer Vision (ICCV), Venice, Italy, 22–29 October 2017; pp. 2223–2232.
12. Ke, X.; Zou, J.; Niu, Y. End-to-End Automatic Image Annotation Based on Deep CNN and Multi-Label Data Augmentation. *IEEE Trans. Multimed.* **2019**, *21*, 2093–2106. [CrossRef]
13. Tran, N.T.; Tran, V.H.; Nguyen, N.B.; Nguyen, T.K.; Cheung, N.M. On data augmentation for GAN training. *IEEE Trans. Image Process.* **2021**, *30*, 1882–1897. [CrossRef] [PubMed]
14. Konidaris, F.; Tagaris, T.; Sdraka, M.; Stafylopatis, A. Generative adversarial networks as an advanced data augmentation technique for MRI data. In Proceedings of the 14th International Joint Conference on Computer Vision, Imaging and Computer Graphics Theory and Applications (VISIGRAPP 2019), Prague, Czech Republic, 25–27 February 2019; Volume 5, pp. 48–59.
15. Liu, B.; Tan, C.; Li, S.; He, J.; Wang, H. A Data Augmentation Method Based on Generative Adversarial Networks for Grape Leaf Disease Identification. *IEEE Access* **2020**, *8*, 102188–102198. [CrossRef]
16. Goodfellow, I.; Pouget-Abadie, J.; Mirza, M.; Xu, B.; Warde-Farley, D.; Ozair, S.; Courville, A.; Bengio, Y. Generative adversarial networks. *arXiv* **2014**, arXiv:1406.2661.
17. Kingma, D.P.; Welling, M. Auto-encoding variational Bayes. *arXiv* **2013**, arXiv:1312.6114.
18. Radford, A.; Metz, L.; Chintala, S. Unsupervised Representation Learning with Deep Convolutional Generative Adversarial Networks. *arXiv* **2015**, arXiv:1511.06434.

19. Chen, X.; Duan, Y.; Houthooft, R.; Schulman, J.; Sutskever, I.; Abbeel, P. InfoGAN: Interpretable Representation Learning by Information Maximizing Generative Adversarial Nets. In Proceedings of the 30th Annual Conference on Neural Information Processing Systems (NIPS 2016), Barcelona, Spain, 5–10 December 2016; pp. 2180–2188.
20. Arjovsky, M.; Chintala, S.; Bottou, L. Wasserstein GAN. *arXiv* **2017**, arXiv:1701.07875.
21. Tolstikhin, I.; Bousquet, O.; Gelly, S.; Schölkopf, B. Wasserstein Auto-Encoders. *arXiv* **2017**, arXiv:1711.01558.
22. Dai, B.; Wipf, D. Diagnosing and Enhancing VAE Models. *arXiv* **2019**, arXiv:1903.05789.
23. Larsen, A.B.L.; Snderby, S.K.; Larochelle, H.; Winther, O. Autoencoding beyond pixels using a learned similarity metric. In Proceedings of the 33rd International Conference on Machine Learning (ICML 2016), New York City, NY, USA, 19–24 June 2016; Volume 4, pp. 2341–2349.
24. Hughes, D.; Salathe, M. An open access repository of images on plant health to enable the development of mobile disease diagnostics. *arXiv* **2015**, arXiv:1511.08060.
25. Szegedy, C.; Liu, W.; Jia, Y.; SErmanet, P.; Reed, S.; Anguelov, D.; Erhan, D.; Vanhoucke, V.; Rabinovich, A. Going deeper with convolutions. In Proceedings of the IEEE Conference on Computer Vision and Pattern Recognition, Boston, MA, USA, 7–12 June 2015; pp. 1–9.
26. Huang, G.; Liu, Z.; Van Der Maaten, L.; Weinberger, K.Q. Densely connected convolutional networks. In Proceedings of the 2017 IEEE Conference on Computer Vision and Pattern Recognition (CVPR), Los Alamitos, CA, USA, 21–26 July 2017; pp. 4700–4708.
27. Heusel, M.; Ramsauer, H.; Unterthiner, T.; Nessler, B.; Hochreiter, S. GANs Trained by a Two Time-Scale Update Rule Converge to a Local Nash Equilibrium. In Proceedings of the 31st Annual Conference on Neural Information Processing Systems (NIPS 2017), Long Beach, CA, USA, 4–9 December 2017; pp. 6629–6640.
28. Salimans, T.; Goodfellow, I.; Zaremba, W.; Cheung, V.; Radford, A.; Chen, X. Improved techniques for training gans. In Proceedings of the 30th Annual Conference on Neural Information Processing Systems (NIPS 2016), Barcelona, Spain, 5–10 December 2016; pp. 2234–2242.
29. Barratt, S.; Sharma, R. A Note on the Inception Score. *arXiv* **2018**, arXiv:1801.01973.

Article

Optimizing the Optimal Planting Period for Potato Based on Different Water-Temperature Year Types in the Agro-Pastoral Ecotone of North China

Jinpeng Yang [1], Yingbin He [1,*], Shanjun Luo [2], Xintian Ma [1], Zhiqiang Li [3], Zeru Lin [3] and Zhiliang Zhang [4]

1   Institute of Agriculture Resources and Regional Planning, Chinese Academy of Agricultural Sciences, Beijing 100081, China; 82101195279@caas.cn (J.Y.); 82101205439@caas.cn (X.M.)
2   School of Remote Sensing and Information Engineering, Wuhan University, Wuhan 430070, China; luoshanjun@whu.edu.cn
3   School of Management, Tianjin Polytechnic University, Tianjin 300387, China; 1930133053@tiangong.edu.cn (Z.L.); 2031131393@tiangong.edu.cn (Z.L.)
4   Institute of Water-Saving Agriculture in Arid Areas of China, Northwest A&F University, Yangling 712100, China; zhiliang_zhang@nwafu.edu.cn
*   Correspondence: heyingbin@caas.cn

Abstract: Potato is the fourth staple crop in China after wheat, maize and rice. The agro-pastoral ecotone (APE) in North China is a main region for potato production. However, potato yield has been seriously constrained by water shortages because of low precipitation and highly variable precipitation patterns during the growing season in this area. In this study, the Agricultural Production Systems Simulator (APSIM) model was used to simulate potato water-limited yield and historical years were divided into different water-temperature year types to optimize the optimal planting period (OPP) and cultivar of potato. The results showed that the potato yield varied in different water-temperature year types. Fast-developing cultivar Favorita could obtain the highest yield in most places and water-temperature year types due to its relatively short length of tuber formation stage. In this study, we suggest changing the planting date according to the water-temperature year type, which offers a new way to adapt to a highly variable climate. However, our method should be adopted carefully because we only considered climate factors; other agronomic management practices (adjusting planting density, plastic film mulch, conservation tillage etc.) also have a great effect on planting date and cultivar selection, which should be further investigated in the future.

Keywords: potato management; tuber formation stage; precipitation patterns

Citation: Yang, J.; He, Y.; Luo, S.; Ma, X.; Li, Z.; Lin, Z.; Zhang, Z. Optimizing the Optimal Planting Period for Potato Based on Different Water-Temperature Year Types in the Agro-Pastoral Ecotone of North China. *Agriculture* **2021**, *11*, 1061. https://doi.org/10.3390/agriculture11111061

Academic Editor: Matt J. Bell

Received: 8 October 2021
Accepted: 25 October 2021
Published: 28 October 2021

Publisher's Note: MDPI stays neutral with regard to jurisdictional claims in published maps and institutional affiliations.

1. Introduction

With the implementation of the "potato as the staple food" strategy, potato has gradually become the fourth staple crop after wheat, corn and rice in China [1]. The APE in North China, characterized by suitable temperature and soil conditions [2–4], is a main region for potato production in China [5]. Rainfed potato is one of the most common crops in this region, accounting for 46.8% of the total crop yield [4]. Therefore, studies on increasing potato production to ensure local food security are meaningful. However, the APE is a sensitive zone to climate change, and highly variable precipitation distributions and amounts, both temporally and spatially, result in large fluctuations in potato yield [4,6]. Thus, to cope with the local highly variable climate and enhance potato production, adaptation strategies must be adopted according to different seasonal precipitation patterns [5,7].

Among many agronomic adaptations, adjusting the planting date and selecting suitable cultivars are two effective way to adapt to local and annual climate variations. Tang et al. [5] conducted several planting experiments and found that the ratio of precipitation to potential evapotranspiration around the potato tuber formation stage accounts for 71% of the potato yield variation in Wuchuan County, a typical site in the middle APE of North China. Li et al. [7] showed that precipitation from the tuber formation stage to maturity could explain 87% of potato yield variation. Yu and Wang [8] showed that rapidly developing cultivars should be planted in a drier place, the eastern APE, according to water consumption of different cultivars and precipitation distribution in critical growth stage. In addition, different maturing cultivars can generate a series of growing season lengths to help the potato's critical growing period better match the local rainy season [9]. Therefore, adjusting the planting date and selecting suitable cultivars to meet the period of the precipitation peak during the potato's critical water stage is a useful method for increasing potato yield [7,10,11]. Li et al. [12] optimized planting date and cultivar of potato in North China using APSIM-potato and suggested late planting should be considered in most locations of North China. Tang et al. [5] divided historical years into wet, dry and normal years according to precipitation and found that the OPP of potato had delayed trends from dry years to wet years in the middle APE. Li et al. [7] also conducted a two-year field experiment to identify the optimal planting date in the middle APE, and his result showed that planting date and cultivar should be selected according to different year types.

Although there have been many studies on optimizing OPP and cultivar of potato in the APE of North China, these studies were constrained in specific site and years. In addition, the effect of different water-temperature year types on optimized planting date and cultivar was not investigated. An integrated study on the development of management strategies (principally planting date and cultivar choice) in different water-temperature year types is particularly essential for improving potato yield and ensuring local food security. Long-term field experiments can help explore the relationship between meteorological factors and the yield of different planting dates; however, experiments such as this are time consuming and relatively expensive [13]. Agricultural system models provide a powerful tool to capture interactions between crop growth and development, agronomic management practices and environmental factors (e.g., planting date × cultivar × location) across multiple seasons [14]. Studies investigating adaptation strategies (e.g., optimizing planting date) to improve crop production using simulation models such as the APSIM have been reported previously [10,11,15–17]. However, no study has focused on optimizing the planting date and selecting suitable cultivars according to different water-temperature year types across the APE in North China. Thus, in this study, we aimed to (1) optimize planting dates and recommend the most suitable cultivar according to different water-temperature year types in the APE of North China; and (2) investigate factors that affect OPP in different water-temperature year types.

2. Materials and Methods

2.1. Study Region, Climate and Soil Data

Twelve agrometeorology experimental sites (AESs) spanning the APE of North China were selected to explore the OPP and suitable cultivar in different water-temperature year types (Figure 1). Climate data including sunshine hours (h), precipitation (mm), daily maximum and minimum daily air temperature (°C) from 1979 to 2019 were obtained from the China Meteorological Administration (http://data.cma.cn, 1 March 2021). Sunshine duration was converted to daily solar radiation using the Angstrom equation with parameters 0.5 for a and 0.25 for b [18,19]. Soil parameters (e.g., soil bulk density, soil organic carbon, soil water pH, drained upper limit, etc.) of each layer were obtained from Han et al. [20], a 10 km resolution global soil profile dataset for crop modeling.

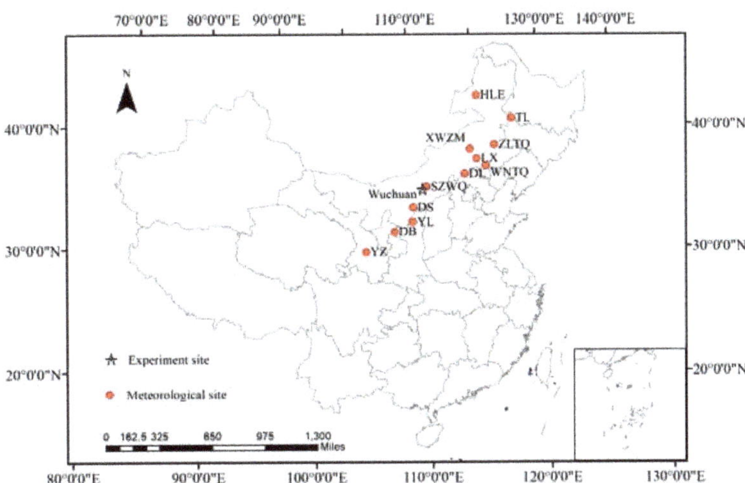

Figure 1. Distribution of experimental site and meteorological sites across the APE in North China. The full name of each meteorological site can be found in Table 1.

Table 1. Study meteorological sites and experiment site.

Site	Latitude (°)	Longitude (°)	Altitude (m)
Dingbian (DB)	37.35	107.35	1360.3
Yulin (YL)	38.16	109.47	1157
Yuzhong (YZ)	35.52	104.09	1874.4
Dongsheng (DS)	39.50	109.59	1461.9
Siziwangqi (SZWQ)	41.32	111.41	1490.1
Duolun (DL)	42.11	116.28	1245.4
Linxi (LX)	43.36	118.04	799.5
Xiwuzhumuqinqi (XWZM)	44.34	117.38	995.9
Wengniuteqi (WNTQ)	42.56	119.01	634.3
Zhaluteqi (ZLTQ)	44.34	120.54	265
Tailai (TL)	46.40	123.45	149.5
Hailaer (HLE)	49.13	119.45	610.2

2.2. Serial Planting Experiments

A series of planting experiments were carried out in 2017 and 2018 in Wuchuan County (111.41° E, 40.49° N, alt. 1756 m), located at the center of the APE in North China (Figure 1). Three different maturing cultivars (fast-developing cultivar Favorita, mid-developing cultivar Connibeck and slow-developing cultivar Kexin_1) were selected to investigate the OPP and most suitable cultivar in different water-temperature year types. The three cultivars have been proven suitable for planting in the APE of North China [11,12]. Potato was planted in 4×7 m plots with three replicates on three planting dates (27 April, 15 May, and 2 June in 2017 and 28 April, 16 May, and 3 June in 2018). The row spacing was 50 cm and the planting density was 5 plants m^{-2}. Urea (46% N), 37.5 kg/hm^2 of potassium chloride and 75 kg/hm^2 of ammonium-diammonium phosphate (18% N) were applied at planting as base fertilizers. Thirty millimeters of irrigation was applied at planting to ensure potato emergence. After that, irrigation was not carried out during the whole potato growing season. All experimental information was obtained from published literature [12].

2.3. APSIM-Potato Model

APSIM-potato is a process-based crop model that has been tested and applied in variable climates across the APE in North China [5,21,22]. It can mimic potato water dynamics, phenology and yield [23]. The key APSIM modules used in our study were potato (simulating potato crop growth and development) and manager (specifying planting, harvest, irrigation and fertilizer rules). Daily weather data (maximum and minimum air temperatures, solar radiation and precipitation), soil properties (e.g., organic carbon content, clay content field saturation capacity, soil lower wilting, etc.) and management information (e.g., planting dates, planting depth, fertilizer, etc.) are needed as inputs to the model. APSIM-potato simulates daily potato growth and development in response to environmental conditions and crop management, including phenological stages, leaf area index, soil water, biomass and tuber yield at a daily time step. The potato phenological phase was divided into six phases from planting to maturity in APSIM potato, i.e., planting-germination, germination-emergence, emergence-early tuber, early tuber-senescing, senescing-senesced and senesced-maturity. Potato requires a certain cumulative thermal time to complete each development stage. The daily dry matter accumulation rate is calculated by radiation interception and radiation-use efficiency and multiplied by water and nitrogen stresses. Potato yield is simulated daily based on partitioning and reallocation of total dry matter to plant organs [24]. Potato genetic parameters of photoperiod for emergence (x_pp_emergence, °Cd), thermal time of planting to emergence (y_tt_emergence, °Cd), emergence to early tuber (tt_earlytuber, °Cd), early tuber to senescing phase (tt_senscing, °Cd) and other parameters were received from published documents [12] (Table 2), which have been well tested in the simulation of potato phenology and yield of different planting dates [12]. To better simulate soil water and thus potato yield, parameters of upper limit of stage 1 evaporation (U) and stage 2 evaporation coefficient which related to soil evaporation were increased to 10 and 4.5 versus default values of 6 and 3.5 due to high evaporation in the APE of China [11].

Table 2. Main potato phenology parameters of APSIM-potato.

Parameter	Favorita	Connibeck	Kexin_1
Degree days from planting to emergence (y-tt-emergence, °C d)	265	320	335
Degree days from emergence to early tuber formation (tt-earlytuber, °C d)	185	205	210
Degree days from early tuber formation to senescing (tt-senescing, °C d)	510	590	660
Photoperiod after emergence (x_pp_emergence, h)		12	
Maximum specific leaf area for delta LAI (y_sla_max, mm^2 g^{-1})		35,000–40,000	

2.4. APSIM Simulation Set up

APSIM 7.10 was used to mimic water-limited potato yield using 41 years (1979–2019) of climate data, and simulations were conducted at 12 locations roughly uniformly distributed across the APE in North China. Planting was simulated at a three-day interval in a potential planting window for three cultivars at each location and each year. The first date of the potential planting window was defined as a five-day running average of daily average temperature higher than 8 °C [4], and the last date was defined when the five-day running average of daily average temperature was lower than 0 °C [11,12]. APSIM was set to harvest when potato matured or when the daily minimum temperature was lower than 0 °C to prevent potato frost events due to the higher risk of frost events in the APE of North China. The OPP of potato was defined by corresponding planting dates exceeding 95% of the peak 15-day running mean water-limited yield in different water-temperature year types at each site.

The crop received 30 mm of irrigation at planting to ensure that it would emerge shortly after it was sown. After that, no irrigation was applied throughout the growth period. APSIM-potato was simulated continuously from 1979 to 2019 with the soil parameters (soil water, soil nitrogen and organic matter, etc.) at the end of last year not resetting, which is more realistic. To avoid nitrogen stress, nitrogen was applied as NO_3^- using a separate fertilizer rule, which was maintained above 300 kg/hm^2 in the top three layers of the soil throughout the whole potato growing season [24].

2.5. Data Processing
2.5.1. Divide Historical Years into Different Water-Temperature Year Types

Forty-one years were divided into different water-temperature year types (dry-cool, dry-hot, wet-cool and wet-hot years) according to the average temperature and the amount of precipitation during the potato growth period in each location. Different water-temperature year types were divided by the following rules (1):

$$\begin{aligned}
&\text{Wet-Hot year: Pre_s} > \text{Pre_a \& Tav_s} > \text{Tav_a} \\
&\text{Wet-Cool year: Pre_s} > \text{Pre_a \& Tav_s} < \text{Tav_a} \\
&\text{Dry-Hot year: Pre_s} < \text{Pre_a \& Tav_s} > \text{Tav_a} \\
&\text{Dry-Cool year: Pre_s} < \text{Pre_a \& Tav_s} < \text{Tav_a}
\end{aligned} \tag{1}$$

where Pre_s is the total precipitation during the potato growing season in a specific year, Pre_a is the average value of Pre_s from historical years, Tav_s is the mean temperature during the potato growth season in a specific year and Tav_a is the average value of Tav_s from historical years.

2.5.2. Statistical Analysis

Linear regression was conducted to test the relationship of potato yield and water and temperature stresses. All statistical analyses and data processing were carried out using the R programming language [25].

The coefficient of variance (CV) was used to represent the year-to-year variation in precipitation in each month:

$$CV = S/X \tag{2}$$

where S is the standard variation for precipitation and X is the mean value of precipitation.

3. Results
3.1. Precipitation Distribution in the APE of North China

The annual precipitation of different months in the potato growing season varied significantly. DL had the lowest CV of 0.43 in June, and LX had the highest CV of 1.29 in October (Table S1). Figure 2 shows an example of the precipitation distribution in different water-temperature year types. The distribution and amount of precipitation varied in water-temperature year types. The variation in precipitation distribution in dry-cool and dry-hot years was relatively lower than that in wet-cool and wet-hot years. The other sites of APE also showed similar features (data not shown).

3.2. Water and Temperature Stresses

The linear regression results showed that water stress in the tuber formation stage explained the most variation in potato yield (Table S2); however, temperature only had a minor effect on potato yield (Table S2, Figure 3). The average water stress in the potato tuber formation stage of different planting dates was lower than 0.2 in dry-hot years and dry-cool years; in contrast, the water stress was lower than 0.3 for most planting dates in wet-hot years and wet-cool years (Figure 3). The average yields in dry-cool and dry-hot years were less than those in cold-wet and warm-wet years. Generally, potato yield increased or decreased with increased or decreased water stress in the tuber formation stage. However, in dry-cool years, potato yield decreased with decreasing water stress when potato was planted after 15 June.

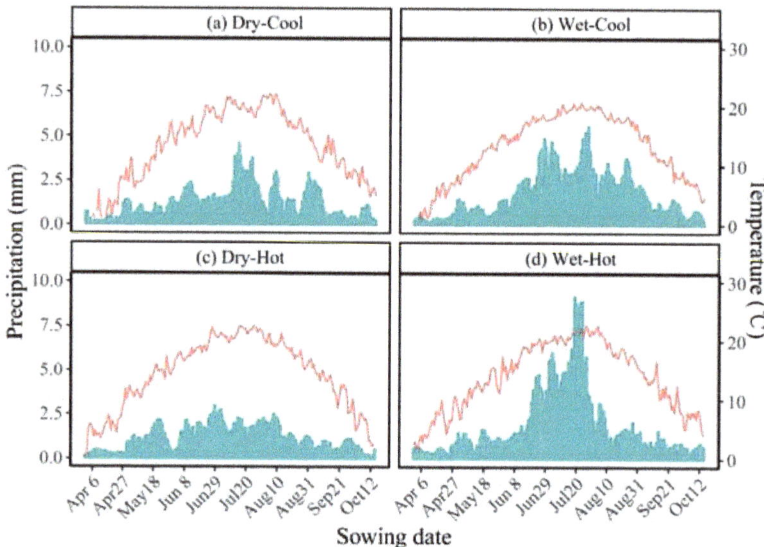

Figure 2. An example of precipitation and temperature distribution during the potato growing season in different water-temperature year types at XWZM of the APE in North China. (**a–d**) represent the dry-cool year, wet-cool year, dry-hot year and wet-hot year, respectively.

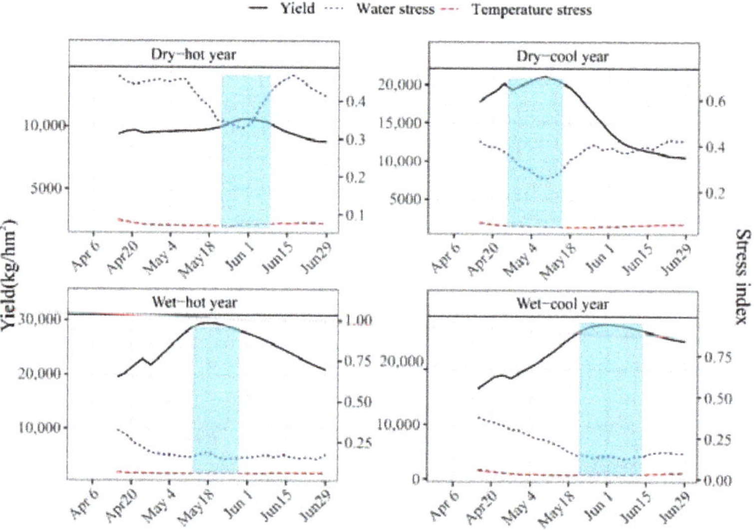

Figure 3. Example of the relationship between average simulated water-limited yield and planting dates for a rapidly developing cultivar (Favorita) in different water-temperature year types at ZLTQ. The blue dashed line indicates the average water stress during the early tuber phase, the red dashed line shows the average temperature stress in the potato growing season, and the cyan rectangle zone represents the optimal planting period at which yield >95% of the running peak mean yield in different water-temperature year types.

3.3. Potato Yield Variation

Potato yield varied in four water-temperature year types across the APE of North China (Figures 4 and S1–S3). The highest yield was achieved by Favorita at TL in the wet-hot year, with an average yield of 31,196 kg/hm². The lowest yield was obtained by Favorita at SZWQ in the dry-hot year, with an average yield of 1882 kg/hm² (Tables S3–S5). Favorita could obtain the highest yield in most locations across the APE in North China in different water-temperature year types. In the dry-cool year, the greatest yield was obtained by the mid-developing cultivar Favorita at DS for planting on 27 June, with an average yield of 21,293 kg/hm². In the dry-hot year, the highest yield was obtained by the rapidly developing cultivar Favorita at TL for planting on 2 May, with an average yield of 17,903 kg/hm² (Figures 4 and S1). In the wet-hot year, the highest yield was obtained by Favorita at TL for planting on May 17, with an average yield of 31,196 kg/hm². In the wet-cool year, the greatest yield was received by Favorita at TL for planting on 29 May, with an average yield of 31,145 kg/hm² (Figures S2 and S3). Potato yield varied across the APE of North China in each type of water-temperature year; however, the trend of potato yields of different cultivars was highly similar. Adjusting the planting date can increase potato yield in each type of water-temperature year; however, in low-yield environments such as SZWQ, potato yield is still very low after the planting date is optimized.

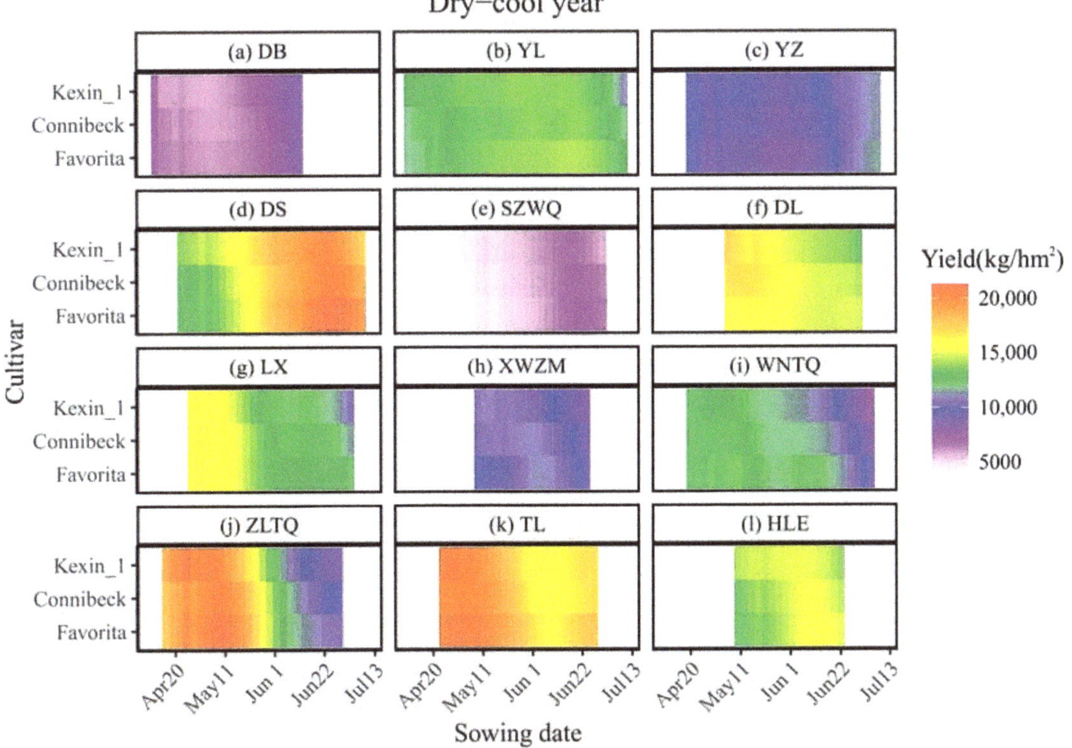

Figure 4. Yield variations across the APE of North China with regard to cultivars and planting dates in dry-cool years. Different horizontal bands represent different cultivars. (**a–l**) refers to different meteorological stations in the APE of North China.

3.4. OPP Variation

The start date of OPP of potato and its duration varied across different water-temperature year types and APE in North China (Figure 5). However, the OPP of different maturing cultivars at each site of APE were similar. The earliest start date of OPP was 9 April, and the latest start date of OPP was 11 July. Both were achieved by Kexin_1 at YL in a dry-cool year (Table S4). The longest duration of OPP was 61 days achieved by the mid-developing cultivar Connibeck at YL in a dry-cool year. The shortest duration of OPP was 4 days, achieved by Favorita and Connibeck at YZ, DB and YL in dry-hot, dry-cool and wet-hot years, respectively (Tables S5 and S6). The average duration of OPP in APE for cultivar Favorita, Connibeck and Kexin_1 are 20.25, 22.25, 21.5 days, respectively, in dry-cool years, 22.25, 21.25, 22.25 days, respectively, in dry-hot years, 23, 25, 26 days, respectively, in wet-hot years and 19, 20, 22 days, respectively, in wet-cool years. Compared to wet-hot and wet-cool years, the distribution of the start date and duration of OPP are more variable in dry-hot and dry-cool years (Figure 5).

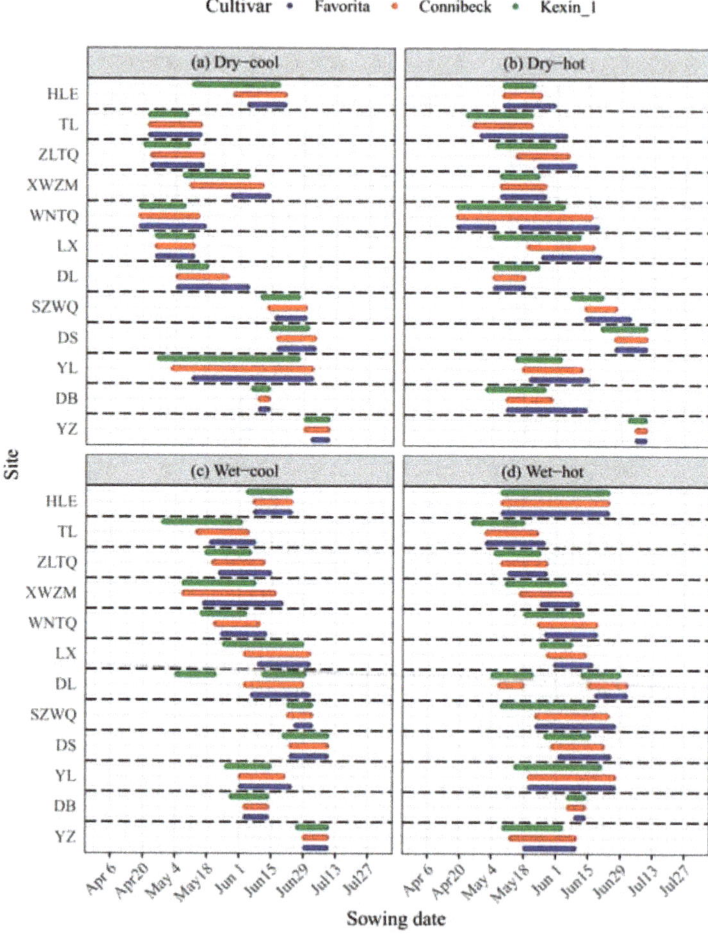

Figure 5. OPP of three mature cultivars in different water-temperature year types across the APE of North China. Different horizontal bands represent different meteorological stations of the APE in North China. (**a–d**) refers to four different year types. Different colors represent different cultivars: blue line for cultivar Favorita, red line for cultivar Connibeck and green line for cultivar Kexin_1.

4. Discussion

4.1. Significance of Dividing Water-Temperature Year Type

The variance in annual precipitation during the potato growth period is very high in the APE of North China. The highest precipitation was 525 mm, while the lowest rainfall was 161 mm, and the coefficient of variation (CV) was 25% [4]. The CV of annual precipitation of different months in the potato growing season also varied significantly (Table S1). To adapt to the highly local variability of precipitation and increase potato yield, Li et al. [12] suggested stabilizing yield by selecting a late planting coupled with mid- and slow-developing cultivars, which is a simple and effective method. The difference with his research is that we advocate changing agricultural management measures to cope with variable climate according to annual precipitation patterns. Therefore, we simulated the potato yield of three different maturing cultivars by using the APSIM-potato model, and divided the historical years (1979–2019) into different water-temperature year types (dry-hot, dry-cool, wet-hot and wet-cool years) across the APE in North China, which is a new attempt to adapt the variability of precipitation distribution in the APE of North China. The results showed that the potato yield and OPP varied in different water-temperature year types across the APE (Figure 2).

The difference in OPP in different seasons indicated the necessity of year-type division. The linear regression results indicated that water stress during the tuber formation stage of potatoes played a dominant role in determining potato yield, while temperature had little effect on yield (Table S2). However, we also found that different water-temperature year types had different precipitation patterns, which means that adding temperature in years dividing provides a method to classify the distribution of precipitation in detail. In actual production and field management, farmers should be told the water-temperature year type in that season by their local Agrometeorological Service, and plant potato at OPP as far as possible and select the most suitable cultivar. This calls for more accurate prediction of future climate [26]. The accuracy of current short-term climate prediction is approximately 70% [27,28], which still cannot fully meet the requirements of agricultural production [29]. Therefore, the identification of OPP will play a more important guiding role in future agricultural production with the improvement of the accuracy of short period and mid-long period climate prediction [7].

4.2. Potato Yield Variation across APE

The APE of North China is a typical arid agricultural area. Many studies have shown that there is a strong correlation between crop yield and precipitation during the growth period under dry land farming [30–33]. However, recent studies have shown that precipitation distribution during crop growth periods has a greater effect on yield [34,35]. The potato tuber formation stage is a critical stage for potato yield and is highly sensitive to water stress [36,37]. A previous study had already shown that there is a good correlation between water conditions in the tuber formation stage and yield in the APE of North China [5]. Our results also showed that the variation in potato yield in different water-temperature year types across the APE in North China is mainly determined by the difference in water stress in the tuber formation stage due to the varied amount and distribution of precipitation. Thus, adjusting the planting date and selecting a suitable cultivar to improve the water condition in the potato tuber formation stage can significantly increase potato yields in the APE of North China (Figures 4 and S1–S3).

Irrigation could significantly increase potato yield in arid and semiarid environments. In recent years, however, the groundwater level in this region has decreased by approximately 0.5–1 m per year to meet the irrigation demand of crops in the APE of North China [38,39] and has induced serious problems of soil salinization and groundwater depletion [40]. In addition to irrigation, plastic film mulching can effectively improve potato yield by reducing soil moisture emissions [41] and increasing soil water storage [42–44]. However, excessive application of plastic film mulching aggravates environmental pollution [45–47]. Adjusting planting dates and selecting suitable cultivars are the simplest and most efficient measures that farmers can choose [48] and impose little pressure on water resources, which is of great significance for ensuring regional food security and sustainable agricultural development. Potato yield can be significantly improved by using an adjusted planting period in different year types. However, other management measures, such as tillage, were not considered in our study, which can also improve potato water productivity and yield [49]. Moreover, in dry-hot and dry-cool years, such as SZWQ, potato yield was still lower than that in wet-hot and wet-cool years after adjusting the planting date and cultivar due to low precipitation (Figures 4 and 5). Tang et al. [10] proposed an effective management strategy to collect rainfall and carry out supplementary irrigation at an appropriate time. Further research should be carried out to study the OPP of potato in combination with supplementary irrigation and cultivar.

4.3. Variation in OPP

Potato yield is closely related to precipitation in the APE of North China, and the starting time of the OPP depends on the starting time of the rainy season [10]. Adjusting the planting date can match the critical water demand stage of potato with the rainy season, thus obtaining sufficient rainfall and minimizing water stress [9]. Our study found that the amount and distribution of precipitation varied in different water-temperature year types (Figure 2), which led to different water conditions in the tuber formation stage corresponding to planting dates and thus resulted in the variance of OPP of potato (Figures 3 and 5). The OPP of different maturing cultivars were very similar; however, the OPP in different water-temperature year types were highly different. Thus, we suggest selecting planting dates according to the water-temperature year type to adapt to various distributions of precipitation.

The duration of OPP represented the variation of the water condition in the tuber formation period of different planting dates. The longer duration of OPP indicates that the variation in rainfall was lower, which resulted in similar water conditions in the tuber formation stages of different planting dates, e.g., YL in dry-cool year and WNTQ in dry-hot year. The shorter OPP indicates that rainfall was concentrated at a certain period, which means that the water stress of the tuber formation stage varies greatly with different planting dates. We further analyzed factors influencing OPP and yield in different water-temperature year types (Figure 3). The OPP of different water-temperature year types varied greatly from April to June, mainly affected by the water stress of the tuber formation stage of different planting dates. However, the yield would be reduced due to the shortening of the growth period when the potato sown too late, especially when a cultivar with a long growth cycle was sown too late (Figure 3). This also explained why the yield of Favorita was slightly higher than that of the other cultivars with a longer growing period.

Li et al. [12] suggested that late planting coupled with fast developing cultivar Favorita was recommended along an 'N-S' transect in North China, while late planting coupled with slow developing cultivar Kexin_1 was recommended along a 'W-E' transect in North China. However, our study found that Favorita can obtain the highest yield in most places of the APE in North China. This is because Favorita can reduce the risk of encountering frost events when planting late due to the relatively short growing period compared to the other two cultivars. Moreover, the duration of the tuber formation stage of Favorita was lower than that of the other two cultivars, which means that Favorita can easily find suitable planting dates with lower water stress in the tuber formation stage (Table 2).

4.4. Uncertainties and Limitations

In this study, we suggest selecting OPP and cultivars according to the water-temperature year type, which responds to the variable distribution of precipitation with changing management practices. However, it is difficult to predict the distribution of precipitation even when the approximate precipitation is known. Therefore, those places in the water-temperature year type with a shorter duration OPP indicate that the distribution of precipitation is relatively similar with historical years, while those places with a longer duration OPP mean that the distribution of precipitation in historical years is more variable, and planting date should be selected carefully in this case. Additionally, readers should note the following limitations in our study. We only considered the impacts of climate factors on potato yield; however, other environmental factors, such as disease, insects and pests, can also impact potato yield and planting date [50,51]. In addition, adjusting planting density, conservation tillage and other agronomic management practices were not considered [12,52], which also have a great impact on planting date and cultivar selection. These influencing factors need to be further investigated in the future. Additionally, to calibrate crop model and further validate our result, more field experiments should be carried out at other sites of the APE in North China in the future.

5. Conclusions

Potato yield and OPP varied in different water-temperature year types; however, the OPP showed little difference between different maturing cultivars. Generally, Favorita obtained the highest yield in different water-temperature year types at most places in the APE of North China. The yield and OPP of potato in different water-temperature year types were mainly affected by water stress in the tuber formation stage due to the varied distribution and amount of precipitation in different water-temperature year types. Compared with unaltered management, increasing yield is recommended by selecting OPP and suitable cultivars according to the water-temperature year type. This study offered a new method to cope with the highly variable climate in the APE of North China, which can help farmers make decisions when climate prediction precision is improved in the future. However, we only considered the impact of climate factors on OPP of potato, but other factors (disease and pests, planting density, conservation tillage, etc.) also have a great effect on OPP and cultivar selection of potato. These factors need to be further explored and more field experiments need to be performed at other sites of the APE in North China in the future.

Supplementary Materials: The following are available online at https://www.mdpi.com/article/10.3390/agriculture11111061/s1, Table S1: Coefficient variation (CV) of year-to-year precipitation in different months of the potato growing season. Table S2: Linear regression between the stress index in the tuber formation stage and potato yield. Table S3: An example of OSP for cultivar Favorita in different water-temperature year types across APE of North China. DOSP represent the duration of OSP. Table S4: An example of OSP for cultivar Connibeck in different water-temperature year types across APE of North China. DOSP represent the duration of OSP. Table S5: An example of OSP for cultivar Kexin_1 in different water-temperature year types across APE of North China. DOSP represent the duration of OSP. Figure S1: Yield variations across APE of North China with regard to cultivars and sowing dates in dry-hot years. Figure S2: Yield variations across APE of North China with regard to cultivars and sowing dates in wet-hot years. Figure S3: Yield variations across APE of North China with regard to cultivars and sowing dates in wet-cool year.

Author Contributions: Conceptualization, J.Y. and Y.H.; methodology, J.Y.; software, J.Y.; validation, J.Y., S.L. and X.M.; formal analysis, Z.L. (Zhiqiang Li); investigation, Z.L. (Zeru Lin); resources, Z.Z.; data curation, J.Y.; writing—original draft preparation, J.Y.; writing—review and editing, J.Y.; visualization, J.Y.; supervision, J.Y.; project administration, Y.H.; funding acquisition, Y.H. All authors have read and agreed to the published version of the manuscript.

Funding: This research was funded by the National Natural Science Foundation of China (41771562) and Project Innovation of Chinese Academy of Agricultural Sciences (IARRP 2021–2025).

Institutional Review Board Statement: Not applicable.

Informed Consent Statement: Not applicable.

Data Availability Statement: The data presented in this study are available on demand from the corresponding author at heyingbin@caas.cn.

Acknowledgments: We would like to thank China Meteorological Administration for providing the historical climate data.

Conflicts of Interest: The authors declare no conflict of interest.

References

1. Lu, X.P. Strategy of potato staple food: Significance, bottlenecks and policy suggestions. *J. Huazhong Agric. Univ.* **2015**, *103*, 1–7.
2. Hijmans, R.J. The effect of climate change on global potato production. *Am. Potato J.* **2003**, *80*, 271–279. [CrossRef]
3. Rykaczewska, K. The Effect of High Temperature Occurring in Subsequent Stages of Plant Development on Potato Yield and Tuber Physiological Defects. *Am. Potato J.* **2015**, *92*, 339–349. [CrossRef]
4. Tang, J.Z.; Wang, J.; He, D.; Huang, M.X.; Pan, Z.H.; Pan, X.B. Comparison of the impacts of climate change on potential productivity of different staple crops in the agro-pastoral ecotone of North China. *J. Meteorol. Res.* **2016**, *30*, 983–997. [CrossRef]
5. Tang, J.Z.; Wang, J.; Fang, Q.X.; Wang, E.L.; Yin, H.; Pan, X.B. Optimizing planting date and supplemental irrigation for potato across the agro-pastoral ecotone in North China. *Eur. J. Agron.* **2018**, *98*, 82–94. [CrossRef]
6. Xia, Z.C.; Pan, Z.H.; Zhang, L.Y.; Zhou, M.M.; Pan, X.B.; Tuo, D.B.; Zhao, P.Y. Examining Mechanisms of Vegetation Ecosystems Degradation Based on Water in Northern Farming-pastoral Zone. *Res. Sci.* **2010**, *32*, 317–322.
7. Li, Y.; Wang, J.; Tang, J.Z.; Ma, X.Q.; Pan, X.B. Selecting planting date and cultivar for high yield and water use efficiency of potato across the agro-pastoral ecotone in North China. *Trans. CSAE* **2020**, *36*, 118–126.
8. Yu, T.T.; Wang, F.X. Planting of different potato cultivars adapted to water availability in Inner Mongolia. *Chin. Agric. Sci. Bull.* **2015**, *31*, 70–77.
9. Liu, K.; Harrison, M.T.; Hunt, J.; Angessa, T.T.; Meinke, H.; Li, C.D.; Tian, X.H.; Zhou, M.X. Identifying optimal planting and flowering periods for barley in Australia: A modelling approach. *Agric. For. Meteorol.* **2019**, *282*, 107871.
10. Hu, Q.; Yang, N.; Pan, F.F.; Pan, X.B.; Wang, X.X.; Yang, P.Y. Adjusting planting Dates Improved Potato Adaptation to Climate Change in Semiarid Region, China. *Sustainability* **2017**, *9*, 615. [CrossRef]
11. Tang, J.Z.; Wang, J.; Wang, E.L.; Yu, Q.; Yin, H.; He, D.; Pan, X.B. Identifying key meteorological factors to yield variation of potato and the optimal planting date in the agro-pastoral ecotone in North China. *Agric. For. Meteorol.* **2018**, *256–257*, 283–291. [CrossRef]
12. Li, Y.; Wang, J.; Tang, J.Z.; Wang, E.L.; Pan, Z.H.; Pan, X.B.; Hu, Q. Optimum planting date and cultivar maturity to optimize potato yield and yield stability in North China. *Field Crop. Res.* **2021**, *269*, 108179. [CrossRef]
13. Singh, R.; Kroes, J.G.; Van Dam, J.C.; Feddes, R.A. Distributed ecohydrological modelling to evaluate the performance of irrigation system in Sirsa district, India: I. Current water management and its productivity. *J. Hydrol.* **2006**, *329*, 692–713. [CrossRef]
14. Padovan, G.; Martre, P.; Semenov, M.A.; Masoni, A.; Bregaglio, S.; Ventrella, D.; Lorite, I.J.; Santos, C.; Bindi, M.; Ferrise, R.; et al. Understanding effects of genotype × environment × planting window interactions for durum wheat in the Mediterranean basin. *Field Crop. Res.* **2020**, *259*.
15. Flohr, B.M.; Hunt, J.R.; Kirkegaard, J.A.; Evans, J.R. Water and temperature stress define the optimal flowering period for wheat in southern-eastern Australia. *Field Crop. Res.* **2017**, *209*, 108–119. [CrossRef]
16. Lilley, J.M.; Flohr, B.M.; Whish, J.P.; Farre, I.; Kirkegaard, J.A. Defining optimal planting and flowering periods for canola in Australia. *Field Crop. Res.* **2019**, *235*, 118–128. [CrossRef]
17. Chen, C.; Fletcher, A.L.; Ota, N.; Flohr, B.M.; Lilley, J.M.; Lawes, R.A. Spatial patterns of estimated optimal flowering period of wheat across the southwest of Western Australia. *Field Crop. Res.* **2020**, *247*, 107710. [CrossRef]
18. Black, J.N.; Bonython, C.W.; Prescott, J.A. Solar radiation and the duration of sunshine. *Q. J. R. Meteorol. Soc.* **1954**, *80*, 231–235. [CrossRef]
19. Jones, H.G. *Plants and Microclimate: A Quantitative Approach to Environmental Plant Physiology*, 2nd ed.; Cambridge University Press: Cambridge, UK, 1992.
20. Han, E.J.; Ines, A.V.; Koo, J. Development of a 10-km resolution global soil profile dataset for crop modeling applications. *Environ. Model. Softw.* **2019**, *119*, 70–83. [CrossRef]
21. Shen, J.J. *Study on Optimalization of Planting Date and Climate Suitability of Main Crops in Agro-Pastoral Ecotone*; China Agricultural University: Pekin, China, 2011.
22. Tang, J.Z. *A Study on Planting Pattern of Potato to Narrow Yield Gap and Increase Precipitation Use Efficiency in the Agro-Pastoral Ecotone*; Agriculture University: Beijing, China, 2019.
23. Chen, R.Y.; Meng, M.L.; Liang, H.Q.; Zhang, J.; Wang, Y.H.; Wang, Z.X. Effects of different treatments of irrigation and fertilization on the yield and nitrogen utilization characteristic of potato. *Chin. Agric. Sci. Bull.* **2012**, *28*, 196–201.
24. Brown, H.E.; Huth, N.; Holzworth, D. A potato model built using the APSIM plant. NET Framework. In Proceedings of the 19th International Congress on Modeling and Simulation, Perth, WA, Australia, 12–16 December 2011; pp. 961–967.
25. R Core Team. *R: A Language and Environment for Statistical Computing*; R Foundation for Statistical Computing: Vienna, Austria, 2016. Available online: https://www.R-project.org/ (accessed on 1 April 2021).
26. Hansen, J.W.; Challinor, A.; Ines, A.; Wheeler, T.; Moron, V. Translating climate forecasts into agricultural terms: Advances and challenges. *Clim. Res.* **2006**, *33*, 27–41. [CrossRef]
27. Jones, J.W.; Hansen, J.W.; Royce, F.S.; Messina, C.D. Potential benefits of climate forecasting to agriculture. *Agric. Ecosyst. Environ.* **2000**, *82*, 169–184. [CrossRef]

28. Zhao, J.H.; Yang, J.; Gong, Z.Q.; Feng, G.L. Analysis of and discussion about dynamic-statistical climate prediction for summer rainfall of 2013 in China. *Adv. Meteorol. Sci. Technol.* **2015**, *5*, 24–28.

29. Fraisse, C.W.; Breuer, N.E.; Zierden, D.; Bellow, J.G.; Paz, J.; Cabrera, V.E.; Garcia, A.G.Y.; Ingram, K.T.; Hatch, U.; Hoogenboom, G. AgClimate: A climate forecast information system for agricultural risk management in the southeastern USA. *Comput. Electron. Agr.* **2006**, *53*, 13–27. [CrossRef]

30. Rockström, J.; Falkenmark, M. Semiarid crop production from a hydrological perspective: Gap between potential and actual yields. *Crit. Rev. Plant. Sci.* **2000**, *19*, 319–346. [CrossRef]

31. Condon, A.G.; Richards, R.A.; Rebetzke, G.J. Breeding for high water-use efficiency. *J. Exp. Bot.* **2004**, *55*, 2447–2460. [CrossRef] [PubMed]

32. Martone, L.; Pilar, P.M.; Grewal, M.S. Long term studies of crop yields with changing rainfall and fertilization. *Agric. Eng. Res.* **2007**, *13*, 37–47.

33. Harms, T.E.; Konscheuh, M.N. Water savings in irrigated potato production by varying hill-furrow or bed-furrow configuration. *Agric. Water Manag.* **2010**, *97*, 1399–1404. [CrossRef]

34. Rockström, J.; Karlberg, L.; Wani, S.P.; Barron, J.; Hatibu, N.; Oweis, T.; Bruggeman, A.; Farahani, J.; Qiang, Z. Managing water in rainfed agriculture: The need for a paradigm shift. *Agric. Water Manag.* **2010**, *97*, 543–550. [CrossRef]

35. Yu, Q.; Li, L.H.; Luo, Q.Y.; Eamus, D.; Xu, S.H.; Chen, C.; Wang, E.L.; Liu, J.D.; Nielsen, D.C. Year patterns of climate impact on wheat yields. *Int. J. Climatol.* **2014**, *34*, 518–528. [CrossRef]

36. Lynch, D.R.; Foroud, N.; Kozub, G.C.; Fames, B.C. The effect of moisture stress at three growth stages on the yield, components of yield and processing quality of eight potato varieties. *Am. Potato J.* **1995**, *72*, 375–386. [CrossRef]

37. Ren, W.J.; Ren, L.; Liu, S.X. Moisture dynamics and supply and demand on potato fields inarid areas of southern Loess Plateau. *Chin. Potato J.* **2015**, *29*, 355–361.

38. Li, D.; Qian, L.H. The estimation of variation of groundwater table in the Inner Mongolia in 2015. *Inn. Mong. Water Res.* **2016**, *5*, 3–5.

39. Zhu, J.L.; Liang, X.J.; Wang, Y.N. Research on the evaluation of groundwater resources in the districts of saving water and increasing grain yield in Jalaid Banner of Inner Mongolia. *China Water Res.* **2014**, *11*, 28–30.

40. Zhang, J.; Zhao, P.Y.; Pan, Z.H.; Duan, Y.; Li, H.C.; Wang, B.; Jing, Y.P.; Dong, Z.Q. Determination of input threshold of nitrogen, fertilizer based on environment-friendly agriculture and maize yield. *Trans. Chin. Soc. Agric. Eng.* **2016**, *32*, 136–143.

41. Gao, S.M.; Zhang, X.C.; Wang, Y.H. Influence of Different Mulching and Furrow-Ridge Planting Motheds on Soil Moisture and Yield of Potato on Dryland. *J. Soil Water Conserv.* **2010**, *24*, 249–251.

42. Hou, X.Q.; Tang, J.; Yu, L.L.; Zhao, F.P.; Wang, Q.W.; Hu, E.J.; Wei, K.R. Effect of autumn mulching tillage on growth and water use efficiency of potato. *J. Drain. Irrig. Mach. Eng.* **2016**, *34*, 165–172.

43. Li, R.; Wang, Y.L.; Wu, P.N.; Sun, R.P.; Qiu, J.X.; Su, M.; Hou, X.Q. Ridge and furrow mulching improving soil water-temperature condition and increasing potato yield in dry-farming areas of south Ningxia. *Trans. Chin. Soc. Agric. Eng.* **2017**, *33*, 168–175.

44. Fan, S.J.; Wang, D.; Zhang, J.L.; Bai, J.P.; Liu, W.X.; Ma, Z.X.; Peng, H.Y. Effects of different cultivation techniques on soil temperature, moisture and potato yield. *Trans. CSAE* **2011**, *27*, 216–221.

45. Liu, E.K.; He, W.Q.; Yan, C.R. 'White revolution' to 'white pollution'—Agricultural plastic film mulch in China. *Environ. Res. Lett.* **2014**, *9*, 091001. [CrossRef]

46. Zhang, D.; Liu, H.B.; Hu, W.L.; Qin, X.H.; Ma, X.W.; Yan, C.R.; Wang, H.Y. The status and distribution characteristics of residual mulching film in Xinjiang. *China J. Integr. Agric.* **2016**, *15*, 2639–2646. [CrossRef]

47. Jiang, R.; Guo, S.; Ma, D.D. Plastic film mulching system and the impact on soil ecological environment in rain-fed drylands of China. *Chin. J. Eco Agric.* **2018**, *26*, 317–328.

48. Zhang, Y.; Zhang, L.Z.; Yang, N.; Huth, N.; Wang, E.L.; Werf, W.V.D.; Evers, J.B.; Wang, Q.; Zhang, D.S.; Wang, R.N.; et al. Optimized planting time windows mitigate climate risks for oats production under cool semi-arid growing conditions. *Agric. For. Meteorol.* **2019**, *266–267*, 184–197. [CrossRef]

49. Hou, X.Q.; Li, R. Effects of mulching with no-tillage on soil physical properties and potato yield in mountain area of southern Ningxia. *Trans. CSAE* **2015**, *31*, 112–119.

50. Sparks, A.H.; Forbes, G.A.; Hijmans, R.J.; Garrett, K.A. Climate change may have limited effect on global risk of potato late blight. *Glob. Chang. Biol.* **2014**, *20*, 3621–3631. [CrossRef] [PubMed]

51. Chandel, R.S.; Pathania, M.; Verma, K.S.; Bhatacharyya, B.; Vashisth, S.; Kumar, V. The ecology and control of potato whitegrubs of India. *Potato Res.* **2015**, *58*, 147–164. [CrossRef]

52. Carter, M.R.; Sanderson, J.B. Influence of conservation tillage and rotation length on potato productivity, tuber disease and soil quality parameters on a fine sandy loam in eastern Canada. *Soil. Tillage Res.* **2001**, *63*, 1–13. [CrossRef]

MDPI

St. Alban-Anlage 66

4052 Basel

Switzerland

Tel. +41 61 683 77 34

Fax +41 61 302 89 18

www.mdpi.com

Agriculture Editorial Office

E-mail: agriculture@mdpi.com

www.mdpi.com/journal/agriculture